Lectures on the Icosahedron and the Solution of the Fifth Degree

FELIX KLEIN

GEORGE GAVIN MORRICE, TRANSLATOR

COSIMO CLASSICS

NEW YORK

Lectures on the Icosahedron and the Solution of the Fifth Degree

For information, address:

Cosimo, P.O. Box 416
Old Chelsea Station
New York, NY 10113-0416

or visit our website at:
www.cosimobooks.com

Lectures on the Icosahedron and the Solution of the Fifth Degree
was originally published in 1884 [date of author's preface].

Cover design by www.kerndesign.net

ISBN: 978-1-60206-306-8

TRANSLATOR'S PREFACE

I HOPE this work may contribute towards supplying the "pressing need of text-books upon the higher branches of mathematic," of which Dr. Glaisher has recently spoken in his presidential address to the London Mathematical Society.

I must here express my deep sense of obligation to Professor Klein for his permission to publish an English translation of his work. The chief difficulty has been to choose proper equivalents for technical expressions in the text, and here Professor Cayley has kindly allowed me to refer to him. I have also consulted Cole's review of the book in Vol. ix. No. 1, of the *American Journal of Mathematics*, and borrowed his expression "self-conjugate subgroup" for "ausgezeichnete untergruppe."

As regards the translation itself, I have been most fortunate in obtaining the help of Miss Borchardt, who, notwithstanding many other engagements, has most kindly gone over most of the manuscript with me.

It would be out of place to speak here of the book itself: a glance at the table of contents will suggest that "tract of beautiful country, seen at first in the distance, but which will bear to be rambled through and studied in every detail of hillside and valley, stream, rock, wood, and flower" (Cayley).

1888.

In preparing the second edition of the translation I have received much valuable help from Professor Burnside, F.R.S., for which I desire to express my warmest thanks.

G. G. M.

October 1913.

PREFACE

The theory of the Icosahedron has during the last few years obtained a place of such importance for nearly all departments of modern analysis, that it seemed expedient to publish a systematic exposition of the same. Should this prove acceptable, I propose to continue in the same course and to treat in a similar manner the subject of Elliptic Modular Functions, and the general investigations newly made of Single-valued Functions, with linear transformations into themselves. Thus a treatise of several volumes would grow, in which I should expect to promote science, at least in so far as it might introduce many to realms of modern mathematics rich in far-stretching vistas.

Referring generally as to the limitations of the material, which I have observed in this publication, to the following exposition itself, I would here only draw attention to the second part, which treats of the solution of equations of the fifth degree. It is now fully twenty-five years since Brioschi, Hermite, and Kronecker in joint labours created the modern theory of equations of the fifth degree. But though these investigations are now and again quoted, the mathematical world at large has hitherto failed to grasp their true import. By giving the first place in the following pages to the subject of the Icosahedron, and by treating this as the true basis of the processes of solution, a view of the theory is brought forward than which a simpler and more lucid one cannot well be desired.

A special difficulty, which presented itself in the execution of my plan, lay in the great variety of mathematical methods entering into the theory of the Icosahedron. On this account it seemed advisable to take for granted no specific previous knowledge in any direction, but rather to introduce, where necessary, such explanations and references as might suffice as preliminary landmarks on the field under immediate survey. What I, however, do expect in my reader is a certain ripeness of mathematical judgment, which shall enable him to interpret concise, brief statements, so as to see in them the general principle involved in the particular case. This is the same method I have ever pursued in my more advanced lectures; indeed I have introduced into the details of these expositions the practices of my lecture-room. It is in this spirit I would have the title interpreted which I have given to my disquisition.

I cannot close these short prefatory remarks without expressing my special thanks to my honoured friends Professor Lie in Christiania and Professor Gordan in Erlangen for manifold suggestions and assistance. My indebtedness to Professor Lie dates back to the years 1869–70, when we were spending the last period of our student-life in Berlin and Paris together in intimate comradeship. At that time we jointly conceived the scheme of investigating geometric or analytic forms susceptible of transformation by means of groups of changes. This purpose has been of directing influence in our subsequent labours, though these may have appeared to lie far asunder. Whilst I primarily directed my attention to groups of discrete operations, and was thus led to the investigation of regular solids and their relations to the theory of equations, Professor Lie attacked the more recondite theory of continued groups of transformations, and therewith of differential equations.

It was in the autumn of 1874 that I first came into real contact with Professor Gordan. I had at that time already commenced the study of the Icosahedron for myself (without then knowing Professor Schwarz's earlier works, to which we shall hereafter frequently have occasion to refer); but I considered my whole manner of attacking the question rather in the light of preliminary training. If now a far-reaching theory has grown from those beginnings, I attribute this result primarily to Professor Gordan. I am not here referring to his trenchant and profound labours, which shall be fully reported upon hereafter. In this place I must record what cannot be expressed in quotations or references, namely, that Professor Gordan has spurred me on when I flagged in my labours, and that he has helped me with the greatest disinterestedness over many difficulties which I should never have overcome alone.

F. KLEIN.

LEIPZIG, *May* 24, 1884.

TABLE OF CONTENTS

PART I

THEORY OF THE ICOSAHEDRON ITSELF

CHAPTER I

THE REGULAR SOLIDS AND THE THEORY OF GROUPS.

CHAPTER II

INTRODUCTION OF $(x+iy)$.

CHAPTER III

Statement and Discussion of the Fundamental Problem, according to the Theory of Functions.

CHAPTER IV

On the Algebraical Character of our Fundamental Problem.

CHAPTER V

General Theorems and Survey of the Subject.

PART II

THEORY OF EQUATIONS OF THE FIFTH DEGREE

CHAPTER I

THE HISTORICAL DEVELOPMENT OF THE THEORY OF EQUATIONS OF THE FIFTH DEGREE.

CHAPTER II

INTRODUCTION OF GEOMETRICAL MATERIAL.

CHAPTER III

The Canonical Equations of the Fifth Degree.

CHAPTER IV

The Problem of the *A*'s and the Jacobian Equations of the Sixth Degree.

CHAPTER V

The General Equation of the Fifth Degree.

PART I

THEORY OF THE ICOSAHEDRON ITSELF

CHAPTER I

THE REGULAR SOLIDS AND THE THEORY OF GROUPS

§ 1. Statement of the Question.

When we speak, in the following pages, of the icosahedron, or in general of a regular solid, this expression is to be understood in an extended sense; namely, we do not actually operate with general constructions in space, but confine ourselves essentially to the *sphere* which is described through the summits of the regular solid, and to which we suppose the edges and sides of the regular solid transferred by linear projection from the centre of the sphere. The nearer object of our consideration is therefore a determinate *partition of the sphere,* and we only return for greater convenience of expression to the phrases, and in part to the constructions, of the geometry of space.

To the regular solids, as the ancients knew them, are usually added in modern times the Kepler solids (whose sides mutually interpenetrate one another). If we wished to transfer them, in the manner explained, by central-projection on to the sphere, a multiple envelopment of the sphere would result. It is, indeed, not hard to see that there is an infinite number of such envelopments of a regular type.* But comparatively complicated relations of this nature shall be set aside in the following pages. We only investigate those simple figures which correspond in the sense mentioned to the regular *tetrahedron,* the *octahedron,* the *cube,* the *icosahedron,* and the *pentagon-dodekahedron.* To these we will then add a sixth configuration, which corresponds to the *plane regular* n-*gon.* In fact, by considering

* *Cf.* in this respect the new work of Hess: "Introduction to the Theory of the Partition of a Sphere, with Special Reference to its Application to the Theory of Equilateral and Equiangular Polyhedra." Leipzig, 1883.

the portion of the plane limited by the sides of the n-gon to be doubled, we can describe this latter as a regular solid—a dihedron, as we will say: only that this solid, contrary to the elementary notion of such, encloses no space. If we transfer the dihedron by central-projection on to the surface of the circumscribed sphere, we have first, corresponding to its n summits, n equidistant points on a great circle (which might be called the equator), between which lie, as projections of the edges of the dihedron, the n pieces into which this circle is divided by the n points. We then, as is natural, make the two half-spheres bounded by the equator to correspond to the two planes which we have just distinguished, and which bound the dihedron.

But—and this must be emphasised from the first—it is not actually the figures themselves here enumerated which, in the following pages, form the subject of our consideration, but rather those *rotations* or *reflexions*, or, shortly, those *elementary geometrical operations* by which the said figures coincide with themselves. *The figures are for us only the framework by means of which we survey the totality of certain rotations or other transformations.* Therefore the individual regular solids will for us be inseparably connected with their *polar figures*, which, like themselves remain unaltered by these operations. In this sense the *octahedron* belongs to the *cube* whose summits correspond to the mid-points of the sides of the octahedron, the *icosahedron* to the *pentagon-dodekahedron*, which has an analogous position.

Starting from the same principle, we will consider with the *tetrahedron* the allied *counter-tetrahedron* (whose summits are diametrically opposite to the summits of the original tetrahedron); we will, finally, in the case of the *dihedron*, mark the two *poles* of the sphere corresponding to the two sides thereof.* There are thus four different forms which lie at the root of our considerations. We will, in what follows, briefly characterise them by the names *dihedron*, *tetrahedron*, *octahedron*, and *icosahedron*. If, in our later developments, we bring the case of the icosahedron in many respects into especial prominence, and if we have, in accordance with this, mentioned the icosahedron

* The configuration of the dihedron is therefore the same as has been elsewhere termed the *double-pyramid*.

alone in the title of this part, it is because the case of the icosahedral configuration is in all respects the most interesting of them.

As soon as we enter upon the task of studying the rotations, &c., in question, by which the configurations which we have mentioned are transformed into themselves, we are compelled to take into account the important and comprehensive theory which has been principally established by the pioneering works of Galois,* and which we term the *group-theory*.† Originally sprung from the theory of equations, and having a corresponding relation with the *permutations* of any kind of elements, this theory includes, as has long been recognised, every question with which we are concerned in the case of a closed manifoldness of any kind of *operations*. We say of any operations that they form a *group*, if any two of the operations, compounded, again produce an operation included amongst those first given. In this sense we have at the outset the proposition:—

The rotations which bring one of the regular solids into coincidence with itself collectively form a group.

For it is clear that any two rotations of this kind, applied one after another, generate again a rotation of the same nature. It is otherwise with the reflexions by means of which a regular solid is transformed into itself. *These taken by themselves in no way form a group.* For two reflexions applied one after another give not a reflexion but a rotation. True, a group will be again formed if we take these reflexions in conjunction with the rotations just mentioned, and certain other operations derived from them by compounding. We shall, moreover, only consider these groups incidentally in the following pages, and shall describe them as the *extended* groups.

* 1829. *Cf.* "Œuvres de Galois" in Liouville's Journal, Series I, tome ii, 1846.

† Though the explanations in the text are limited almost entirely to considerations of the theory of groups, the geometer will be interested, apart from these, in the remarkable relations of position which arise in an individual case on the basis of group-theory properties, and are governed by them. I should like to call attention here to the researches which Herr Reye and M. Stephanos have devoted in this sense to the theory of the cube ("Acta Math.," t. i, p. 93, 97; "Math. Ann.," xxii, p. 348, 1883).

§ 2. Preliminary Notions of the Group-Theory.

Before we turn to the special groups which occur in connection with the regular solids, it will be useful to make mention of certain general notions which have been worked out in other directions in the theory of groups. I beg the reader, who is as yet not conversant with these theories, to make himself acquainted, in conjunction with the short exposition which is to be given here (and which on later occasions will be further completed in various directions), with one of the more detailed expositions * which the theory of groups has lately received. In what follows we consider, with certain exceptions, only finite groups. Such a group is first characterised by the number N, of the operations which it embraces, where we always count the so-called "identical" operation as one; we denote this number as the *order* of the group. Further, we shall give the *periodicity* of the individual operations, *i.e.*, the number of repetitions which the individual operation needs in order to return to identity; and, moreover, we shall give the totality of the *sub-groups*, *i.e.*, all such combinations of a part of our operations as, taken together, possess the character of a group. The degree of a sub-group is always a factor of the degree N of the principal group. The simplest sub-group (and we may say generally the simplest group) is always that which arises from the repetitions of an individual operation, whose order, therefore, is equal to the period of the operation in question; such may be called respectively *cyclic* sub-groups and groups.

But a mere enumeration of the things here required is not sufficient; we desire rather to take cognisance of the position of the several operations, sub-groups, and so on, within the main group. In this connection let us consider the following

* *Cf.* J. A. Serret, "Traité d'algèbre supérieure" (Paris, 4th edition, 1879), in German by Wertheim (Leipzig, 2nd edition, 1878–79); C. Jordan, "Traité des substitutions et des équations algébriques" (Paris, 1870); E. Netto, "Substitution theorie und ihre Anwendung auf die Algebra" (Leipzig, 1882). Particularly should reference be made to the articles which Herr Dyck has published in the 20th and 22nd volumes of the "Mathematische Annalen" (1882–83), as "Gruppentheoretische Studien."

definitions. Let us agree first that by the product of two operations S and T:

$$ST$$

we will understand that operation which consists in first allowing S and then T to operate. In general

$$ST \text{ is not} = TS.$$

If this occurs in a special case, we call the two operations S and T *permutable*. We construct now in general

$$STS^{-1} = T',*$$

(where S^{-1} denotes that operation which compounded with S produces 1, *i.e.*, identity). If S and T are not permutable, T' is different from T; we say, then, that T' proceeds from T by transformation, and call T and T' *associates* within the main group. In fact, T' will correspond with T in all essential properties, *e.g.* (as we see at once), it has the same periodicity.

Now let T be replaced by the operations T_1, T_2 . . . T_k . . . of any sub-group. Then the same thing happens (as we apply each time the same S to every T) to the corresponding T', so that, in fact, $T'_i T'_k = T'_e$, when $T_i T_k$ coincides with T_e.† We say that the groups of T and T' are then themselves conjugates within the main group.

We must consider now in particular the case where two different sub-groups (the original and the transformed) coincide with one another. If this occurs in the case of a set of operations, which we may choose from the entire group for the transformation of our sub-group, and if our sub-group thus shows itself only associated with itself, then we call it a *self-conjugate* sub-group. Every group contains, if we like to press the definition so far, two self-conjugate sub-groups: viz., in the first place, the totality of all its operations, *i.e.*, the group itself, and, in the second place, that simplest group which consists of the identical operation alone. If a group contains, apart from these improper cases, no self-conjugate sub-groups, it is called *irreducible*, otherwise it is called *composite*.

In the case of composite groups we seek especially their *decomposition*. We effect a decomposition of a group by giving

* If $T' = STS^{-1}$, $(T')^2 = STS^{-1} \cdot STS^{-1} = ST^2S^{-1}$; generally $(T')^r = ST^rS^{-1}$. If, then, $T^n = 1$, $(T')^n = 1$ also and conversely, *q.e.d.*

† For we have again $T'_i T'_k = ST_iS^{-1} \cdot ST_kS^{-1} = ST_iT_kS^{-1} = ST_eS^{-1}$.

a self-conjugate sub-group, as extensive * as possible, contained in it; then, again, a new one, self-conjugate within the sub-group so obtained, &c., and so on till identity is reached. It need hardly be said that this decomposition process admits of much variation according to circumstances.

Beyond these simplest definitions, which come under our notice in the case of individual groups, I must consider that relation between two groups which is described as *isomorphism*. Two groups are called isomorphic if their operations can be so exhibited that $S_i S_k$ always corresponds to $S'_i S'_k$ provided that S_i is made to correspond to S'_i and S_k to S'_k.

The isomorphic relation can be a mutually unique one; we then speak of simple isomorphism. In this case the two groups, from an abstract point of view, are in general identical, and it is only in the *significance* of the two sets of operations that a difference can exist. The sub-groups of the one group, therefore, give directly the sub-groups of the other group, &c., &c.

But the co-ordination may also be an ambiguous one, and then we describe the isomorphism as *multiple*. Here again to every sub-group of the S group corresponds one of the S' group, and *vice versa*, but the two sub-groups need not possess the same degree. At the same time, associate sub-groups of the one give similar sub-groups of the other. Therefore, also, self-conjugate sub-groups of the one group are transformed into similar ones in the other. In particular to identity, if we attribute it to the S group, corresponds a self-conjugate sub-group within the S' group and conversely.†

In what follows we shall have principally to do with examples of multiple isomorphism, in which to each S corresponds one S', but to every S' two S's are co-ordinated (so that the number of the S's is double as great as the number of the S''s). We shall then simply speak of hemihedric isomorphism.

§ 3. The Cyclic Rotation Groups.

Turning now to the closer consideration of the groups which are formed by the rotations which bring one of the configura-

* That is, one which is not contained in a sub-group more comprehensive and at the same time self-conjugate.

† *Cf.* besides the publications already mentioned, in particular: "Capelli, sopra l'isomorfismo . . ." in Bd. 16 of "Giornale di Matematiche" (1878).

tions mentioned in § 1 into coincidence with itself, we must give precedence to the simplest rotation groups, *those which are obtained by the repetition of a single periodical rotation.* Evidently, for such a group, two points on our sphere remain unaltered, which points we will call the two *poles;* and the group consists, if it contains on the whole n rotations, of the n rotations through an angle

$$=0,\ \frac{2\pi}{n},\ \frac{4\pi}{n},\ \ldots\ \frac{2(n-1)\pi}{n}$$

round the axis joining the two poles.

We agree in the first place that any two rotations of this group are permutable with one another. Therefore every individual rotation, as well as every sub-group which can be composed with individual rotations, is only associated with itself. But whether such sub-groups exist depends on the character of the number n. If n is a prime number, the existence of a proper sub-group is *a priori* excluded (because its degree must be a factor of n); if n is composite, there is, corresponding to every factor of n, one and only one sub-group whose degree is equal to this factor.* We shall obtain a *decomposition* of our group if we first seek the sub-group which in this sense corresponds to the highest factor contained in n, and then further treat the sub-group thus obtained in the same way.

If we like to familiarise ourselves with the idea of isomorphism here directly, we observe that our group is simply isomorphic with the totality of the "cyclic" permutations of any n elements taken in a definite order:

$$(a_0, a_1, a_2, \ldots a_{n-1}).$$

In fact, we can establish a correspondence between the permutations alluded to and the rotations which we have been considering most simply by geometrical means. We have only to construct the *n points:*

$$a_0, a_1, a_2, \ldots a_{n-1}$$

which are derived from an arbitrarily given point a_0 by our rotations, and now remark how these points are permuted amongst themselves by the rotations.

* I make these and similar statements in the text without proof, because they will either be self-evident to the reader, or must be apparent to him, without further proof, on a little reflection.

It is superfluous to spend more time over such obvious matters. We had to introduce them because the cyclic groups are, so to say, the elements from which all others are constructed.

§ 4. The Group of the Dihedral Rotations.

Turning now to the configuration of the dihedron, I beg the reader—here and in the similar developments of the following paragraphs—to make the corresponding diagrams, or to think out for himself directly by aid of a model—which is easily constructed—the properties under consideration. For we are treating of concrete matters, which may easily be conceived with the assistance of the suggested aids, but which may occasionally offer difficulties if these are neglected. I should also have had throughout to lay down these developments much more in detail, had I not wished to take for granted the reader's co-operation in the manner explained.

We have already named that great circle on our sphere which carries the n summits of the dihedron the *equator*, and have also already marked the two corresponding *poles*. Then it is clear from the first that the dihedron is transformed into itself by the cyclic group of n rotations for which these poles remain unchanged. But the group of the rotations belonging to the dihedron is not thereby exhausted. We will mark a new point on the equator, midway between some two consecutive dihedral points; the points so obtained we call the *mid-edge points* of the dihedron. We then further describe that diameter which contains a summit or a mid-edge point of the dihedron as a secondary axis thereof. There are n secondary axes of the dihedron; if n is odd, each of these contains one summit and one mid-edge point; if n is even, the secondary axes separate into two categories, according as they connect two summits or two mid-edge points. In every case the dihedron remains unaltered *if it is turned right round on any one of these secondary axes: i.e.*, if it is rotated through an angle π round the secondary axis. Thus, by the side of the cyclic group of n rotations already explained, there are arrayed n other rotations, each of the period 2.

Besides the rotations here enumerated the dihedron group con-

tains no others. In fact, we recognise in the following way (which will be again applied later on) that the number of the dihedral rotations must be equal to $2n$. We consider, first, that every point of the dihedron can be transformed into every other point by means of a dihedral rotation, which admits of n possibilities, and then that, while we keep one summit fixed, the dihedron can only be brought into coincidence with itself in two ways, viz., by a revolution on the secondary axis, which passes through the summit in question, and by the identical operation. Now the number of dihedral rotations must evidently be equal to the *product* of the two factors; it will therefore be equal to $2n$, *q.e.d.*

I will not now weary the reader by enumerating all the sub-groups contained in the dihedral group. Let us rather consider forthwith our first cyclic group of n rotations, and prove *that this, as a sub-group within the main group of the dihedron, is self-conjugate.* In fact, let us go back to the definition of § 2. We denote by T, T', rotations round the principal axis of the dihedron, and by S any other dihedral rotation. Then our assertion requires us to show that $STS^{-1}=T'$. But if S itself denotes a rotation round the principal axis, this relation is self-evident; and if S is a revolution round a secondary axis, then the effect of this revolution, so far as the principal axis is concerned, will be reversed by the operator S^{-1} following, from which our relation again results.

We can refer the proof here given to a general principle, which we introduce here the more readily because in the sequel it will be repeatedly applied. Let us agree first that we will describe in our configurations all such geometrical figures as proceed from one another by an operation of the corresponding group as *conjugates.* We now construct all figures which are associate with a given one. Let T_i be those operations of our group which have the property of leaving unaltered every one of the figures so constructed. *Then the T_i evidently form a self-conjugate sub-group within the main group.* For every operation ST_iS^{-1} belongs itself to the T_i, because S only effects a permutation of the fundamental figures, which will be reversed by S^{-1}. The application of this principle to our case is clear. We have only to consider the two poles of the dihedron as fundamental associate figures. It is here incidental (as far as the general principle is concerned) that those rotations, which leave one of these poles unaltered, do not in general differ from those which transform both poles together into themselves.

By similar reflexions we determine those among the dihedral rotations which are associated with one another. I say with regard to this, that now, *of the rotations round the principal axis, those two which rotate through* $\frac{2k\pi}{n}$ *and* $-\frac{2k\pi}{n}$ *are conjugates, while the revolutions round the secondary axes, for n uneven, are all associates, but, for n even, separate into two categories of conjugates.* The first statement corresponds to the circumstance that the two poles of the dihedron are respectively equally affected by the two rotations round the principal axis which we are comparing,* the latter statement to the earlier theorem that the secondary axes of the dihedron are either all associates, or, for n even, divide into two sorts of associated lines. And further, in both cases we apply a general principle, which we can express by saying: *Those two operations are always conjugates which transform respectively two associated figures analogously into themselves.* I do not spend more time over the proof of this principle.

If, finally, a decomposition of the dihedral group is required, such an one is already implicitly contained in what has gone before. As a sub-group at once the most comprehensive and self-conjugate, we choose the group of n rotations round the principal axis. This we treat further in accordance with the theorems of the preceding paragraph. We define another group of permutations of letters which is simply isomorphic with the dihedral group. For this purpose we will now denote the n summits of the dihedron in their natural order by

$$a_0, a_1, \ldots . a_{n-1}.$$

Then we have first, as in the preceding paragraph, corresponding to the n rotations round the principal axis, those cyclic permutations of the a_v's which replace respectively a_v by a_{v+k} (the indices being taken for the modulus n). We find, further,

* Inasmuch as a rotation through $-\frac{2k\pi}{n}$ round one pole coincides with a rotation through $+\frac{nk}{2\pi}$ round the other.

that by a revolution round the axis which passes through the point a_0, a_v will be replaced by a_{n-v}. From both operations together springs the *metacyclic* group,* which will be represented by the following transformation of indices;

k (mod. n),

isomorphic with our dihedral
ning—is identical with it in an

DRATIC GROUP.

egoing paragraph, as also the de-
assume that n is >2. If $n=2$,
its definite character, inasmuch
ron can then be connected by
ircles. In accordance with this
e, as the corresponding group
uous † group. Interesting and
eory of the continuous groups is
little moment in the following
n the case of $n=2$, make the
e by selecting from among the
rcles passing through the two
s equator. The principal axis of
he two secondary an orthogonal
ct accordance with the rules of
corresponding group of $2n=4$
ual determination of co-ordinates
d, the point x, y, z will be trans-
the other points:

Kronecker every group of permutations
$'\equiv cn+k$ (mod. n).
s of *Lie* in the *Norwegischen Archiv* (from
ath. Ann." Latterly, *M. Poincaré*, in his
n quote) of single-valued functions with
ves, uses the word "continuous group"
such every group of infinitely many but
initely small transformations occur. This
s, however, to me to be not to the purpose.

$$x, \ -y, \ -z;$$
$$-x, \ +y, \ -z;$$
$$-x, \ -y, \ z.$$

Clearly our new group contains, apart from identity, only operations of period 2, and it is incidental that we have connected one of these operations with the principal axis of the figure, and the other two with the secondary axes. So I will give the group a special name which no longer recalls the dihedral configuration, and call it the *quadratic group.* The quadratic group has the special property, as is at once proved, that all its operations are permutable.* Thus every operation appears as only associated with itself.† We shall effect the decomposition of the quadratic group by first descending to an arbitrary sub-group of 2 rotations, for which one of the three axes remains fixed, and then passing from this to identity.

§ 6. The Group of the Tetrahedral Rotations.

We remarked above that, for all rotations which bring a regular tetrahedron into coincidence with itself, the counter-tetrahedron will also be transformed into itself. These tetrahedra, by their eight summits, together determine a *cube.* If we now mark those 6 points on the sphere which correspond to the middle points of the sides, we obtain the 6 summits of a regular *octahedron.* We thus recognise already the close relation in which the group of the tetrahedral rotations stands to the octahedral group which we are now going to study. We will complete our figure by adding thereto the rectangular triad of the diagonals of the octahedron, and also the 4 cube-diagonals (passing through the centre of the sphere).

Applying now the principles developed in § 4, we find, first, *that the tetrahedral group contains* 12 *rotations.* In fact, there are 4 associate tetrahedral points, and each of

* It is easily shown that two rotations are only permutable if either (as in the case of the quadratic group) their axes cross at right angles and each has the period 2, or (as in the case of the cyclic group) their axes coincide.

† This is not contradicted by the fact that, *in the more comprehensive group to be now studied,* the 3 rotations of the period 2, which the quadratic group contains, appear as equivalent.

these summits remains unchanged by 3 rotations—by the identical rotation, and by two rotations of the period 3 whose axes are respectively the cube-diagonals which passes through the tetrahedral point. We have ascertained at once, by what has been said, *that* 8 *of our* 12 *rotations possess the period* 3. Of these (again in virtue of the principles enunciated in § 4) four are conjugates, namely, all those sets of four which appear to rotate *in the same sense*, through an angle of $\frac{2\pi}{3}$ $\left(\text{or } \frac{4\pi}{3}\right)$, round the summit of the tetrahedron which they leave fixed. To these 8 rotations and identity are then added 3 *more associated rotations of period* 2. These are revolutions round the 3 mutually ~~rectangular~~ diagonals of the octahedron, which latter now appear as mutually conjugate, because they are interchanged by each rotation of period 3. Together with identity, the 3 rotations in question evidently form a quadratic group.

We conclude at once that the quadratic group so obtained is self-conjugate within the tetrahedral group. For the 3 mutually associated diagonals of the octahedron all remain unaltered for the rotations of the quadratic group, and only for them. We can therefore decompose the group of the tetrahedron by first descending to the quadratic group, and then, treating this further in the sense of the preceding paragraph; I omit the proof that any other decomposition of the tetrahedral group is not possible, and that generally, except the quadratic group, there exist within the tetrahedral group no sub-groups other than the simple cyclic groups which arise from the repetition of a single rotation.*

Let us consider, further, the nature and manner of the permutations which the 4 diagonals of the cube (which we will shortly denote by 1, 2, 3, 4) undergo in virtue of the tetrahedral rotations. First, we have the self-evident assertion that by no one of the tetrahedral rotations (apart from identity) are all the 4 diagonals of the cube left unaltered.

* Theoretically speaking, we generate all the sub-groups of a given group by first constructing all the cyclic group mentioned in the text, and then combining these with one another in sets of two, three, &c., in order. In each individual case such a process can of course be considerably shortened by appropriate considerations.

There are, therefore, no 2 tetrahedral rotations which generate the same permutation of the four diagonals of the cube. *Therefore the group of the tetrahedral rotations is simply isomorphic with the group of the corresponding permutations of the diagonals of the cube.**

We see, in particular, that to the rotations of the self-conjugate quadratic group correspond the following arrangements of the 4 diagonals:

1,	2,	3,	4;
2,	1,	4,	3;
3,	4,	1,	2;
4,	3,	2,	1;

To these are added, if we proceed to the remainder of the tetrahedral rotations, 8 more which arise from cyclic permutations of 3 out of the 4 diagonals. We have thus, as we see, obtained just those 12 permutations of the 4 diagonals which we are accustomed to call the even permutations.

§7. The Group of the Octahedral Rotations.

In the case of the group of the octahedral rotations, we have, as has been already pointed out, essentially the same configuration for a foundation as in the case of the tetrahedron. We will only further mark (on our sphere) the 12 points which correspond to the mid-edge points of the octahedron, and construct the 6 diameters which contain a pair of these points. These 6 diameters we call tho cross-lines of the figure.

Of course the octahedral group contains the 12 rotations of the tetrahedral group, and indeed, as we can premise, as a self-conjugate sub-group. For the 8 summits of the cube admit of being distributed between the tetrahedron and counter-tetrahedron in only one way, and these latter remain both unaltered by the twelve rotations in question. In

* Let us compare the behaviour of the 3 octrahedral diagonals. These, since they remain unaltered for the operations of the quadratic group, are permuted by the 12 tetrahedral rotations only in 3 ways, viz., cyclically. With the group formed of these permutations, the tetrahedral group is then *multiply* isomorphic.

addition to these, 12 *more rotations then arise which interchange the tetrahedron and counter-tetrahedron, so that the octahedral group contains on the whole* 24 *rotations.* These are: first, 6 rotations —mutually conjugate—through an angle π round the 6 cross-lines of the figure, then 6 rotations through $\frac{\pm\pi}{2}$ (therefore of period 4) round the 3 diagonals of the octahedron. The latter prove themselves also mutually conjugate. For the 4 rotations, for which the individual diagonals of the octahedron remain unmoved, now participate as a self-conjugate sub-group, in a dihedral group of 8 rotations. Similarly the two rotations of period 3 round the same diagonal of the cube, and therefore in general all rotations of the period 3, are associate. For every diagonal of the cube is principal axis of a dihedral group of 6 rotations. The rotations of period 2, on the other hand, separate into two sharply defined categories, according as a diagonal of the octahedron or a cross-line remains fixed by them. The decomposition of the octahedral group is formed of course by descending first to the tetrahedral group, and then to the quadratic group, &c., &c. No other kind of decomposition exists, as we have now exhausted in advance all the sub-groups contained in the octahedral group.

Finally, we agree that the diagonals 1, 2, 3, 4, of the cube, in virtue of the 24 rotations of the octahedral group, are permutated in 24 ways. *The octahedral group is, therefore, simply isomorphic with the totality of the permutations of* 4 *elements.*

§ 8. The Group of the Icosahedral Rotations.

The group of the icosahedron, to which we now turn, is for us the most interesting of them all, because, as we shall show, it is *primitive*, in contradistinction to the groups of the dihedron, tetrahedron, and octahedron. It shares this property with those cyclic rotation groups whose order is a prime number.

For the sake of investigating the group of the icosahedron, let us imagine that (in addition to the 12 icosahedral points), as we have said already, the 20 summits of the corresponding pentagon-dodecahedron (which correspond to the middle points of the sides) are constructed on the surface of our sphere, and, further, the 30 points which correspond on the sphere to the mid-edge points of the icosahedron. The 12 icosahedral points distribute themselves in pairs on 6 diameters, which we will describe shortly as *diagonals* of the icosahedron. Similarly, corresponding to the 20 summits of the pentagon-dodecahedron, we speak of 10 *diagonals* of the pentagon-dodecahedron, and finally of 15 *cross-lines* containing by pairs the mid-edge points.

We convince ourselves first *that the total number of icosahedral rotations is* 60. In fact, each of the 12 (evidently mutually conjugate) icosahedral points remains unaltered on the whole by 5 rotations. We have thus at once (of course leaving the identical substitution out of the question), corresponding to each of the 6 diagonals of the icosahedron, 4 rotations of the period 5, in general, therefore, 24 rotations of this kind. In the same sense the 10 diagonals of the pentagon-dodecahedron give $10 \cdot 2 = 20$ rotations of period 3, and the 15 cross-lines 15 rotations of period 2, whereby if we add identity the totality of the 60 rotations is exhausted:

$$24 + 20 + 15 + 1 = 60.$$

Of the rotations here enumerated the 15 of period 2, and similarly the 20 of period 3, prove themselves respectively associate; for the 15 cross-lines and the 10 diagonals of the pentagon-dodecahedron are so, and, if we rotate round one of these diagonals through $\frac{2\pi}{3}$ or $\frac{4\pi}{3}$, it comes to the same, so far as the main group is concerned, as if the two end points were again associated. On the grounds of similar considerations the rotations of period 5 are separated into two categories of 12 associates. The first category contains all rotations which turn through an angle of $\pm \frac{2\pi}{5}$ round one of the diagonals of the icosahedron, the other those whose angle of rotation amounts to $\pm \frac{4\pi}{5}$.

With these data we have at once determined the *cyclic subgroups* which are contained in the icosahedral group. There are, as we see, 15 such groups having $n=2$, 10 groups having

$n=3$, 6 groups having $n=5$; cyclic groups with the same n are always conjugate.

These data are sufficient to prove that the icosahedral group is simple. Namely, if a self-conjugate sub-group existed, this would have to contain either all or none of the cyclic groups having $n=2$ (because these are associates), so too of the cyclic groups $n=3$ or $n=5$, either all or none. But the groups $n=2$, 3, 5, bring with them respectively 15, 20, 24 operations different from identity. If therefore we denote by η, η', η'', three numbers which can represent 0 or 1, the number of operations contained in the assumed self-conjugate sub-groups amounts to:

$$1+15\,.\,\eta+20\,.\,\eta'+24\,.\,\eta''.$$

But now this number, as we remarked before, must be a factor of the degree of the main group, and therefore of 60; this necessarily gives either:

$$\eta=\eta'=\eta''=0,$$

whereby our sub-group coincides with identity, or:

$$\eta=\eta'=\eta''=1,$$

which means that the sub-group is not distinct from the main group. *The icosahedral group is therefore simple, q.e.d.*

Next to the cyclic sub-groups we find in the case of the icosahedron, as a glance at the model teaches us, for further sub-groups, first, 6 conjugate dihedral groups having $n=5$ and 10 associate dihedral groups having $n=3$. The former have the diagonals of the icosahedron, the latter those of the pentagon-dodecahedron as principal axes; the corresponding secondary axes are contributed by the 15 cross-lines. We might suppose that in a similar manner, corresponding to the 15 cross-lines, 15 dihedral groups would present themselves having $n=2$, *i.e.*, quadratic groups. Here, however, arises the fact that in the case of the quadratic group the principal axis is equivalent with the two secondary axes. *In correspondence with this we obtain only* 5 *mutually conjugate quadratic groups.* These correspond one by one to the 5 rectangular triads into which we can divide the 15 cross-lines.

In these quadratic groups we have encountered that property

of the icosahedron which in the following pages will interest us most of all. Inasmuch as only 5 rectangular triads, as we have remarked, can be formed out of the 15 cross-lines, each of these triads must remain unaltered, not only by the rotations of the corresponding quadratic group, but on the whole for 12 icosahedral rotations. *It can be shown that these rotations form a tetrahedral group.* In fact, the 8 summits of the cube which corresponds to the rectangular triad to be considered are all included in the 20 summits of the pentagon-dodecahedron.* There are, therefore, contained *eo ipso* among the icosahedral groups those 8 rotations of period 3, which, together with the rotations of the fundamental quadratic group, form a tetrahedral group. We will also expressly agree that the 5 tetrahedral groups so formed are associates.

Leaving again the proof that, besides those enumerated, no other sub-groups of the icosahedral group exist, let us only further observe the isomorphism which arises in the case of the icosahedral group from the existence of the aforesaid 5 rectangular triads. It can be shown that for every rotation of period 5 these triads are cyclically interchanged in a definite order. For each rotation of period 3, on the other hand, 2 of the triads remain unaltered, and only the other 3 are interchanged in cycle. Finally, it appears that for every rotation of period 2 one of the triads remains unaltered, while the other 4 are interchanged in pairs. *In this manner it is shown that the group of the* 60 *icosahedral rotations is simply isomorphic with the group of* 60 *even permutations of* 5 *things.*

We could of course here, as in former cases, have exhibited the essential isomorphism of our groups with certain groups of permutations of symbols, and then have transferred to the former the results which are found in the text-books with regard to the latter groups. Now we have investigated our groups directly, *i.e.*, by means of the figures themselves, it will be a useful exercise to compare the results obtained by us with the known properties of isomorphic groups.

* One sees occasionally (in old collections) models of 5 cubes, which intersect one another in such a way that their $5 . 8 = 40$ summits coincide in pairs, and represent the 20 summits of a pentagon-dodecahedron.

§ 9. On the Planes of Symmetry in our Configurations.

For the further progress of our developments it is useful to construct the appropriate *planes of symmetry* of our configurations, *i.e.*, those planes with respect to which the configuration is its own reflexion, and then consider the partition of the sphere which is effected by these planes.

In the case of the dihedron, we can construct, besides the plane of the equator, n other planes of symmetry, viz., those planes which contain, besides the principal axis, one of the secondary axes. By means of these $(n+1)$ planes, the sphere will be cut up into $4n$ congruent isosceles triangles, which have 2 angles $=\frac{\pi}{2}$ and one angle $=\frac{\pi}{n}$. Of such triangles 4 meet in each dihedral point and in each mid-edge point, and $2n$ in each of the two poles, at equal angles.

In the case of the regular *tetrahedron*, there exist 6 planes of symmetry, viz., those planes which, passing through an edge of the tetrahedron, are at right angles to the opposite edge. Consider for a moment a tetrahedron proper, limited by 4 planes, situated in space. Clearly each of the 4 equilateral triangles in these planes will be cut up by the planes of symmetry, through the agency of its 3 perpendiculars, into 6 alternately congruent and symmetric triangles. If we now transfer this partition by central projection on to the sphere, we have on the latter 24 alternately congruent and symmetric triangles, of which each exhibits the angles, $\frac{\pi}{3}, \frac{\pi}{3}, \frac{\pi}{2}$, and which at the summits of the original tetrahedron, as also at the summits of the counter-tetrahedron, meet in sets of 6, at the summits of the corresponding octahedron in sets of 4, with angles respectively equal.

In the case of the *regular octahedron*, in addition to the planes of symmetry of the tetrahedron, which as such are retained, 3 more arise: those planes which contain 2 of the 3 diagonals of the octahedron. By the 9 planes thus obtained, the surface of the octahedron (which we will suppose for a moment to be a solid proper, constructed independently in space), consisting of 8 equilateral triangles, will be partitioned in a manner just like that in which the surface of the tetrahedron has been. Passing by central projection to the sphere, we obtain thereon 48 alternately congruent and symmetric triangles with the angles $\frac{\pi}{3}, \frac{\pi}{4}, \frac{\pi}{2}$, which meet at the summits of the octahedron in sets of 8, and at the ends of the cross-lines (the mid-edge point of the octahedron) in sets of 4. This is that partition of the sphere which is well known in crystallography in the case of the so-called Achtundvierzigflächner. Finally, in the case of the icosahedron, we have, as planes of symmetry, those 15 planes which contain two of the six diagonals of the icosahedron. These partition the 20 equilateral triangles which are contained in the bounding surfaces of the icosahedron considered as a solid, exactly in the manner now several times considered. We obtain, therefore, 120 alternately congruent and symmetric triangles on the sphere, whose angles are $\frac{\pi}{3}, \frac{\pi}{5}, \frac{\pi}{2}$, and which meet in the summits of the pentagon-dodecahedron in sets of 6, in the summits of the icosahedron in sets of 10, and in the ends of the cross-lines in sets of 4. Let us consider the similarity of the results thus obtained in the four cases. In each we have to do with a partition of the sphere into alternately congruent and symmetric triangles * which meet in sets of 2ν in those points of the spherical surface which remain unmoved by a cyclic sub-group of ν rotations. Of the numbers ν there are in every case three, corresponding to the summits of the several triangles. They appear, in order of magnitude, collected in the following table, which may be kept in view during the later developments:

	ν_1	ν_2	ν_3
Dihedron . .	2	2	n
Tetrahedron . .	2	3	3
Octahedron . .	2	3	4
Icosahedron . .	2	3	5

* When we speak above, in the case of the dihedron, shortly of congruent triangles, no absurdity is involved, for we can, even in this case, describe the triangles as alternately congruent and symmetric, inasmuch as it is with isosceles triangles that we have to do.

We observe at once that the number of the triangles is in every case double as great as the degree of the corresponding group of rotations (which we will in future denote by N); they amount in the four cases respectively to $4n$, 24, 48, 120.

We complete these developments further by constructing, in the case of *the cyclic groups* also, certain planes which we call their symmetry-planes. These are to be simply such n planes, passing through the corresponding pole, as proceed from one another by means of the rotations of the group. These planes decompose the sphere into $2n$ congruent (or, if we prefer, alternately congruent and symmetric) *lunes*, of angular separation $\frac{\pi}{n}$, of which each extends from one pole to the other.

§ 10. General Groups of Points—Fundamental Domains.

We now apply the spherical partitions which we have obtained to the closer study of our groups of operations. We consider, first, the groups of points which arise if we submit an arbitrary point to the N rotations of our group, and which we will call the *aggregate of points or group of points belonging to our group of operations.* Here we will suppose, for the sake of a clearer representation and more convenient description, the bounded regions on the sphere to be alternately *shaded* and *not shaded.* It is manifest *a priori that for the rotations of a single group each shaded region will be transformed once, and only once, into every other shaded region, and similarly each non-shaded region once, and only once, into each non-shaded region.* In fact, the number N of the rotations, as already remarked, coincides in every case with the half of the total number of regions.

If, now, any point on the sphere is given (which may belong either to a shaded or non-shaded region), we can, in virtue of our space-partition, without further trouble give the $(N-1)$ new positions which it assumes by means of the $(N-1)$ rotations—distinct from identity—of our group; we have simply to mark those $(N-1)$ points which are situated within the $(N-1)$ remaining shaded or non-shaded regions, in just the same way as the initial point in the original region. In general the N points of the group of points so arising are all distinct; they only partially coincide when the initial point retires to a *summit* of the surrounding region. If, on the whole, ν shaded (and of course the same number of non-shaded) regions meet at this summit, then the point will remain unaltered by ν rotations of the group, and only assume on the whole $\frac{N}{\nu}$ different positions. The *special* sub-groups of points so arising are none other than those which we have otherwise considered in the foregoing paragraphs in our investigations of the individual groups.*

With the groups of points here constructed is connected a conception which will later on be of use to us. *We describe as the fundamental domain of a group of point-transformations in general such a portion of space as contains one, and only one, point of every corresponding group of points.*† The boundary points of such a domain are connected naturally in pairs by means of the transformations of the group, and only half of them can be attributed to it. I say now *that, for our groups, we may consider as a fundamental domain in any case the combination of a shaded and a non-shaded region.* In fact, if we allow a point to traverse a region thus defined without crossing its own track, the corresponding group of points cover uniquely the whole surface of the sphere.

§ 11. The Extended Groups.

Applying the suggestions of § 1, we now *extend* the groups hitherto considered, by connecting with their rotations the reflexions of the respective configurations on the planes of symmetry.

Here also the partition of the sphere given in § 10 will be of service to us. In fact, we recognise at once *that each several region there distinguished, shaded or non-shaded, is a fundamental domain of the extended group, and that therefore the*

* For the general groups of points mentioned in the text, consult the work of *Hess* already alluded to, where they are used for the purposes of the theory of polyhedra.

† *Cf.* for different uses of this notion (so important for all applications of the theory of groups to geometry) my "New Contributions to Riemann's Theory of Functions," in the xxi. Bd. Math. Annalen (1882).

extended group contains just $2N$ *operations.* As regards the proof of this statement, let us note, first, that a combination of the rotations hitherto considered with the reflexion on a single plane of symmetry suffices for turning each of our shaded regions into each of the non-shaded regions. On the other hand, let us reflect that a deformation of the sphere which is known to be a rotation, or to spring from a combination of a rotation with a reflexion, is completely determined as soon as we know that it transforms one of our regions into a definite one elsewhere.

The fundamental domains so obtained have, in contradistinction to those considered in the foregoing paragraphs, the peculiarity of being in no wise arbitrary. In fact, their boundary points are *a priori* determined by the fact that each remains unaltered by a determinate operation of the extended group, viz., by reflexion on a plane of symmetry. We can generate the extended group by connecting the initial group of rotations with the reflexion on that particular plane of symmetry in which the boundary point under consideration is contained. *Therefore the special groups of only* N *points, which spring from the boundary points of the fundamental domains by the application of the extended group, are at the same time general groups of points in the sense of the foregoing paragraphs.* Moreover, they are the only ones amongst these groups of points which at the same time remain unaltered by the operations of the extended group. Of course the special point-groups of $\frac{N}{\nu}$ points just mentioned, corresponding to the summits of the fundamental domains, are also included among them.

We might here have investigated our new groups, *the extended groups*, in the same sense by the theory of groups, as we have done for the original groups in the preceding paragraphs. I should like to recommend such a discussion to the reader as an appropriate exercise, and limit myself here in this direction to the following statement:—The original group is in every case manifestly self-conjugate within the extended group. But, besides this, the extended octahedral and icosahedral groups, as well as the extended dihedral group for n even contain a self-conjugate sub-group of only two operations.

This springs from a double application of that transformation which replaces every point on the sphere by that diametrically opposite to it.*

§ 12. Generation of the Icosahedral Group.

Hitherto, in our consideration of groups, we have supposed the individual groups ready to hand, and sought to obtain a uniform view of their different operations, and of the position of these latter with regard to one another. In the following pages, however, we shall find a more one-sided process of practical value. Our business will be to introduce the groups by *appropriate generating operations, i.e.*, to present operations from which, by repetition and combination, the group in question arises.

We treat, first, in this sense the group of the icosahedral rotations, here again taking advantage of the partitioning of § 9 and the fundamental domains of § 10 respectively. The principle, which here serves as our basis, has been already implicitly applied in the preceding paragraphs. Since each fundamental domain of a group will only be obtained from any other by *one* operation of the group, we can *name* the different fundamental domains after the operations, in virtue of which they proceed from an arbitrary one amongst them, which we will denote by 1, as being the initial domain. Effecting this nomenclature, we obtain directly from it an enumeration of all the operations of the group.†

We will suppose, for the sake of a more convenient mode of expression, that the icosahedron is so placed that one of its diagonals runs vertically. For a first fundamental domain we then choose one of the 5 isosceles triangles, which, endowed

* As especially remarkable, I will add that the extended octahedral group, consisting of 48 operations, contains 3 different self-conjugate sub-groups of 24 operations. These are first, as is manifest, the original octahedral group and the extended tetrahedral group, and then that group which consists in a combination of the original tetrahedral group with the operation just mentioned in the text. Only the latter group, not the "extended" tetrahedral group, is a sub-group of the "extended" icosahedral group.

† Consult here the already mentioned "Gruppentheoretischen Studien" of Herr Dyck, in Bd. xx. of Math. Ann. The principle mentioned in the text is there applied to the general purposes of the theory of groups.

with angles $\frac{2\pi}{5}, \frac{\pi}{3}, \frac{\pi}{3}$, are grouped on the sphere round the uppermost summit of the icosahedron: such a triangle is a fundamental domain of the icosahedron group, because it is composed of two neighbouring triangles of the partitioning given in § 9. The five isosceles triangles in question form, we will say, a first *pentagon* of the pentagon-dodecahedron belonging to the icosahedron. Those sides of a triangle which are at the same time *sides of a pentagon* we will describe as the *ground-lines* of the figure in question.

We now denote by S the rotation in a determinate direction through an angle, $\frac{2\pi}{5}$, round the vertical diagonal of the icosahedron. Thus the 5 fundamental domains before mentioned will proceed in their natural order from the first of them by the rotations:

$$1,\ S,\ S^2,\ S^3,\ S^4,$$

we will therefore denote the domains by the symbols S^μ, $\mu = 0, 1, 2, 3, 4$.

We now take a second icosahedral rotation, T, of period 2. This shall be the revolution round that cross-line of the icosahedron, one of whose ends is the mid-edge point of the ground-line of 1. By means of this T, our 5 domains S^μ are transformed into the domains $S^\mu T$, which, taken together, again make up a pentagon of our pentagon-dodecahedron, and, in fact, that one which has in common with the first pentagon just considered the ground-line of the first fundamental domain. Applying now again the operations S, S^2, S^3, S^4, we obtain from the new pentagon the remaining 4 attached to the first pentagon. Therefore, the fundamental domains of those 5 pentagons which surround the first one are represented by:

$$S^\mu T S^\nu,\ (\mu,\ \nu = 0, 1, 2, 3, 4).$$

A third icosahedral rotation, also of period 2, shall now be denoted by U, of which, however, we shall see that it has no independent importance, but is compounded of the two S and T. The axis of U shall coincide with one of the cross-lines which run horizontally, and, indeed, to make everything determinate, we will choose that horizontal cross-line in particular which stands at right angles to T. Clearly the rotation U so determined transforms the 6 upper pentagons of the pentagon-dodecahedron which we have hitherto considered into its 6 lower pentagons, which were still wanting. Therefore we find at once *that the thirty fundamental domains of the icosahedral group which were still wanting are given by the following:*

$$S^{\mu}U,\ S^{\mu}TS^{\nu}U,\ (\mu,\ \nu=0,\ 1,\ 2,\ 3,\ 4).$$

From the fundamental domains we now turn back to the rotations. We then have the proposition, the deduction of which was the object of our present considerations, viz., *that the* 60 *rotations of the icosahedral group which are given by the following scheme:*

$$S^{\mu},\ S^{\mu}TS^{\nu},\ S^{\mu}U,\ S^{\mu}TS^{\nu}U,\ (\mu,\ \nu=0,\ 1,\ 2,\ 3,\ 4).$$

Here the rotations:

$$S^{\mu},\ S^{\mu}U$$

form the dihedral group $n=5$ belonging to the vertical diagonal of the icosahedron, and the rotations:

$$T,\ U,\ TU$$

give, when taken together with identity, one of the 5 quadratic groups occurring in connection with the icosahedron.

If we draw a figure, as seems indispensable for the full understanding of the theorems here developed, or if we operate, as is more convenient, by means of a model of the icosahedron on which the different fundamental domains are marked out, and the corresponding symbols introduced, we can of course at once read off all the operations which make up *any* sub-group of the icosahedral group. We have only to mark those fundamental domains which proceed from the domain 1 by the operations of the sub-group.*

It remains for us to generate U, as we proposed, by a combination of S and T. To this end we subject, say, the fundamental domain S^3TS^2 to the operation T. Thus arises

* *E.g.*, I find for the tetrahedral group which embraces the quadratic group just noted:

$$1,\ T,\ STS^3,\ S^3TS,\ S^2TS^4,\ S^4TS^2$$
$$U,\ TU,\ STS^3U,\ S^3TSU,\ S^2TS^4U,\ S^4TS^2U.$$

a fundamental domain S^3TS^2T which belongs to one of the pentagons of the lower half. But we have previously called this same domain (as a glance at the figure shows) TS^3U. Hence:

$$S^3TS^2T = TS^3U.$$

In this equation let us consider U as the unknown. We solve this equation by first multiplying by T on both sides of the left-hand expression and then by S^2 and recalling that $T^2 = 1$, $S^5 = 1$. In this manner we have:

$$U = S^2TS^3TS^2T,$$

and this is the relation we wanted.

§ 13. Generation of the other Groups of Rotations.

As regards the generation of the other groups of rotations, this can follow without further trouble by the same means as we have now applied to the case of the icosahedron. But for the first of these, the cyclic and dihedral groups, the matter is so simple that we need no special method, and for the tetrahedron and octahedron we propose in the sequel to use a method of generation which runs parallel with the decomposition of these groups before noted. I gather together here the results in question, which are easy to verify without special deduction.

Now, as regards the cyclic groups, their operations will be manifestly given by the symbols:

$$S^\mu, \Big(\mu = 0,\ 1,\ 2,\ \ldots\ (n-1)\Big),$$

where S denotes the rotation through the angle $\frac{2\pi}{n}$. We obtain the group of the *dihedron* if we annex any revolution T round one of the auxiliary axes of the dihedron, and therefore add to the operations S^μ the others:

$$S^\mu T, \Big(\mu = 0,\ 1,\ 2,\ \ldots\ (n-1)\Big).$$

In particular, the operations of the *quadratic group* are now represented (in agreement with the data just given) by the following scheme:

$$1,\ S,\ T,\ ST.$$

From the quadratic group we now ascend to the tetrahedral group by annexing any one of the corresponding rotations of period 3, which we will call U. The 12 rotations of the tetrahedron will then be given by the following table:

$$\begin{array}{llll} 1, & S, & T, & ST, \\ U, & SU, & TU, & STU, \\ U^2, & SU^2, & TU^2, & STU^2. \end{array}$$

Finally, we get the 24 rotations of the octahedral group on annexing to the 12 rotations here enumerated the others:

$$\begin{array}{llll} V, & SV, & TV, & STV, \\ UV, & SUV, & TUV, & STUV, \\ U^2V, & SU^2V, & TU^2V, & STU^2V. \end{array}$$

Here V denotes *any one* octahedral rotation which is not contained in the tetrahedral group, *e.g.*, a rotation of period 4 round one of the octahedral diagonals.

We here conclude these preliminary considerations. Their object was to introduce into comparatively elementary geometrical figures the ideas of the theory of groups, in such a form that the group-theory reflexions and the geometrical mode of illustration might henceforward supplement one another.

CHAPTER II

INTRODUCTION OF $(x+iy)$

§ 1. First Presentation and Survey of the Developments of this Chapter

The essential step for our further progress in developing our train of thought is as follows: to consider the sphere which we submitted to the groups of rotations, &c., and on which we studied the corresponding groups of points and fundamental domains, as now the vehicle of the values of a complex variable $z=x+iy$. This method of representation, originating with *Riemann*, and first thoroughly expounded by *Herr C. Neumann* * in his "Vorlesungen über Riemann's Theorie der Abel'schen Integrale," is at the present day sufficiently well known, so that I can make use of it immediately; besides, the formulæ furnished in the following paragraphs are in themselves an efficient introduction to the theory.

In virtue of the representation thus introduced, the individual system of points which we have hitherto considered appears defined by an *algebraical equation* $f(z)=0$, where the degree of f is identical with the number of points, as long as none of these points retires to $z=\infty$, which declares itself, in the well-known manner, by a fall of one unit in its degree. We inquire what properties these equations possess corresponding to the circumstance that the groups of points represented by them are transformed into themselves by certain rotations of the sphere, or by certain reflexions, &c.

* Leipzig, 1865. *Cf.* for the general application of Riemann's method my treatise, "Ueber Riemann's Theorie der algebraischen Functionen und ihrer Integrale" (Leipzig, 1882). *Cf.* again for the connection of this introduction of $(x+iy)$ with the projective treatment of surfaces of the second order my work (to which I shall have further occasion to allude), "Ueber binäre Formen mit linearen Transformationen in sich selbst," in Bd. 9 of Math. Ann. (1875), particularly at p. 189.

With regard to this we have, first, the fundamental theorem, which I will presently establish and define more precisely, viz., *that every rotation of the* $(x+iy)$ *sphere on its centre will be represented by a linear substitution of* z*:*

$$(1) \qquad z' = \frac{\alpha z + \beta}{\gamma z + \delta}.$$

In fact, the z, which we can suppose extended with its complex value over the original sphere, and the z', which, in just the same way, we can suppose extended over the rotated sphere, are, in virtue of the interdependence of the two different spheres, related to one another *uniquely* without exception; and, moreover, since the relation between the two spheres is one of conformity,* they are *analytically* related to one another; they are, therefore, by known theorems, linearly dependent on one another.† So, too, we recognise that, to the reflexions and other inverse operations (which spring from the composition of a reflexion with arbitrary rotations), correspond formulæ of the following kind:

$$(2) \qquad z' = \frac{\alpha \bar{z} + \beta}{\gamma \bar{z} + \delta}.$$

where $\bar{z}$ denotes the conjugate imaginary value $(x-iy)$ of z. *Our equations* $f(z)=0$ *have, therefore, the property of remaining unaltered by a group of linear substitution* (1), *or, in some cases, by an extended group which contains, alongside of substitutions* (1), *a corresponding number of substitutions* (2).‡

* It is indeed one of *congruency*, since the corresponding points of either sphere can be brought into coincidence with one another by rotation.

† Unfortunately we find the fundamental theorems of the function-theory, such as we are now considering, developed in the text-books in such a form that the conformable figure which is furnished by the functions is only incidentally taken into consideration; it is, therefore, for our purpose necessary to make, in every case, a certain modification and combination of the proofs explicitly given; these, however, can present no difficulty to the reader, since we are always concerned with quite elementary relations.

‡ The same, of course, is true of equations $F(z)=0$, which, when combined, represent *several* groups of points such as are considered in the text. We can consider these equations $F(z)=0$ as a generalisation of the *reciprocal* equations of lower analysis, inasmuch as the latter also remain unaltered by a definite group of linear substitutions, viz., by the simple group $z'=z$, $z'=\frac{1}{z}$.

I must now consider at once the analytical method which occurs spontaneously in the establishment of the equations $f(z)=0$ and in the study of their mutual relations, and which, by virtue of its more varied aspects, excels in many respects the former reflexions based on geometrical illustrations: that of *the homogeneous variables.* If we replace z by $z_1 : z_2$, the substitution (1), (and analogously every substitution (2)), splits up into two separate operations:

$$(3) \qquad \begin{aligned} z'_1 &= \alpha z_1 + \beta z_2, \\ z'_2 &= \gamma z_1 + \delta z_2, \end{aligned}$$

where now the absolute value of the determinant $(\alpha\delta - \beta\gamma)$ of the substitution will be of especial importance. Instead of the equations $f(z)=0$ or $f\left(\frac{z_1}{z_2}, 1\right)=0$, we shall then have to consider *the form* $f(z_1, z_2)$, on multiplying by a proper power of z_2. This form has always the same degree (a first point in favour of the homogeneous notation) as the corresponding group of points, the occurrence of the point $z=\infty$ being now indicated by a factor z_2 of f. We recognise at the same time that, with the transition to a *form* f, a new distinction arises. For f need not remain absolutely unaltered for the substitutions (3); it can change to a factor près, and our business will be to determine this factor. Moreover, we obtain, by putting in the foreground the consideration of the theory of forms, a bond of union with that important theory of modern algebra which is described as the *theory of invariants of binary forms.* This will be of service to us in the more complicated cases, in order to deduce from one form f all the rest in a simple manner. I may mention at once the result in which the considerations here explained culminate (see the paragraph before the last of this chapter). It is this; *that for each group of linear substitutions* (1) *corresponding to our group of rotations, a corresponding rational function:*

$$(4) \qquad Z = R(z),$$

will be found, which represents the different groups of points belonging to the group, if we equate it to a parametric constant. But at the same time we obtain, if we actually represent those groups of substitutions, a series of new problems, from which, later on, our further development will have to start.*

§ 2. On those Linear Transformations of $(x+iy)$ which Correspond to Rotations Round the Centre.

Let the equation of our sphere, relatively to a system of rectangular central co-ordinates, be:

$$(5) \qquad \xi^2+\eta^2+\zeta^2=1.$$

We then introduce the complex magnitude $z=x+iy$, by first exhibiting $(x+iy)$ in the usual manner in the $\xi\eta$-plane (the equatorial plane), and then, placing this plane by stereographic projection from the pole $\xi=0$, $\eta=0$, $\zeta=1$, in a (1, 1) relation with the surface of the sphere. We thus obtain the formulæ:

$$(6) \qquad x=\frac{\xi}{1-\zeta},\ y=\frac{\eta}{1-\zeta},\ x+iy=\frac{\xi+i\eta}{1-\zeta};$$

or:

$$(7) \qquad \xi=\frac{2x}{1+x^2+y^2},\ \eta=\frac{2y}{1+x^2+y^2},\ \zeta=\frac{-1+x^2+y^2}{1+x^2+y^2}.$$

As we particularly want to determine these linear substitutions of z which correspond to the rotations of the sphere, the *diametral* points of the sphere are of interest to us (inasmuch as one pair remains unmoved in every rotation). In order to derive, with reference to these, a preliminary theorem, we substitute in (6), instead of ξ, η, ζ, their negative values. Then we have for the diametral point:

$$x'-iy'=\frac{-\xi+i\eta}{1+\zeta},$$

and therefore, by multiplication with the values (6) of $(x+iy)$ and attending to (5):

* Consult throughout the work already mentioned, "Ueber binäre Formen mit linearen Transformationen in sich selbst," in Bd. 9 of Math. Ann. (1875). It is there that for the first time that process of thought is displayed from its foundations which now reaches a detailed exposition in the developments of the first and second chapters of the text. I had communicated the principal results in June 1874 to the Erlanger physikalisch-medicinische Gesellschaft (*cf.* the Sitzungsberichte).

(8) $$(x+iy)(x'-iy')=-1,$$

or, if we put $(x+iy)=re^{i\phi}$,

(9) $$x'+iy'=\frac{1}{r}\cdot e^{i(\phi+\pi)}.$$

Diametrically opposite points have arguments whose absolute values are reciprocal, while their amplitudes differ by π.

We consider now, first, the case where the axis $0-\infty$ (which stands at right angles to the plane of the equator) is rotated through an angle a, and let this rotation, looking from the outside on to the point ∞ (which we suppose placed on the upper side of the equatorial plane), take place in a sense opposite to that of the hands of a clock. A point, which originally had the argument of z, will, after the rotation, have the argument z'. We inquire how z' is connected with z. Evidently in the same way as $(\xi'+i\eta')$ with $(\xi+i\eta)$, if we rotate the $\xi\eta$-plane (the equatorial plane) in the way given; for the denominator $(1-\zeta)$ in the formulæ (6) remains unaltered by the rotation. But now we have for the said rotation of the $\xi\eta$-plane, if, as usual, we let the positive ξ-axis extend to the right, and the positive η-axis away from us:

$$\xi'=\xi\,.\cos a-\eta\,.\sin a,$$
$$\eta'=\xi\,.\sin a+\eta\,.\cos a,$$

or,

$$\xi'+i\eta'=(\cos a+i\sin a)\,(\xi+i\eta);$$

whence follows in the well-known manner:

(10) $$z'=e^{ia}\,.\,z.$$

If we now wish to represent analogously a rotation through an angle a, for which the points ξ, η, ζ, and $-\xi$, $-\eta$, $-\zeta$, on the sphere remain unmoved, and for which the first point plays the same part as the point ∞ did before—so that, therefore, if we view ξ, η, ζ, from *without*, the rotation takes place in a sense opposite to that of the hands of a clock—we have in (10), instead of z and z', such a linear function of z and z' respectively as becomes infinite at ξ, η, ζ, and vanishes at $-\xi$, $-\eta$, $-\zeta$. Such a linear function is, however, determined, save as to a factor; it runs in its most general form:

$$C\frac{z+\dfrac{\xi+i\eta}{1+\zeta}}{z-\dfrac{\xi+i\eta}{1-\zeta}};$$

But it is unnecessary to determine this factor more precisely by any kind of convention, because it must of itself drop out of the formula to be established. In fact, we obtain, on substituting in (10) for z our expression, independently of C:

$$\frac{z'+\dfrac{\xi+i\eta}{1+\zeta}}{z'-\dfrac{\xi+i\eta}{1-\zeta}}=e^{i\alpha}\cdot\frac{z+\dfrac{\xi+i\eta}{1+\zeta}}{z+\dfrac{\xi+i\eta}{1-\zeta}}$$

or, after an easy transposition alteration:

$$(11)\qquad e^{\frac{-i\alpha}{2}}\cdot\frac{z'(1+\zeta)+(\xi+i\eta)}{z'(1-\zeta)-(\xi+i\eta)}=e^{\frac{i\alpha}{2}}\cdot\frac{z(1+\zeta)+(\xi+i\eta)}{z(1-\zeta)-(\xi+i\eta)}.$$

This is therefore the general formula for an arbitrary rotation, for which we sought. If we solve it for z', it will be convenient to introduce the following abbreviations:

$$(12)\qquad \xi\sin\frac{\alpha}{2}=a,\quad \eta\sin\frac{\alpha}{2}=b,\quad \zeta\sin\frac{\alpha}{2}=c,\quad \cos\frac{\alpha}{2}=d,$$

where evidently:

$$(13)\qquad a^2+b^2+c^2+d^2=1.$$

We then obtain the simple form:

$$(14)\qquad z'=\frac{(d+ic)\,z-(b-ia)}{(b+ia)\,z+(d-ic)}.^{*}$$

We have, as we might suppose *a priori*, obtained by this method *two formulæ* for every rotation of the sphere. The rotation remains unaltered, namely, if we increase the angle of rotation α by 2π. Now the consequence of this is, by formula (12), that all 4 magnitudes change their sign. This corresponds to the circumstance that the determinant of the substitution of (14) will be equal to $a^2+b^2+c^2+d^2$, therefore by (13) equal

* See the note by Cayley in Bd. 15 of Math. Ann. (1879), "On the Correspondence of Homographies and Rotations," where this formula is for the first time explicitly established.

to 1, which in respect to the signs of a, b, c, d, admits of just two possibilities.

At the same time we have obtained a convenient rule for calculating the cosine of half the angle of a rotation which is given in the form:

$$z' = \frac{Az+B}{Cz+D},$$

and thereby estimating the periodicity of this substitution (so far as we are concerned with periodic substitutions). For manifestly we have, by comparison with (14):

$$\text{(15)} \qquad \cos\frac{\alpha}{2} = \frac{A+D}{\sqrt{AD-BC}}.$$

§ 3. Homogeneous Linear Substitutions—Their Composition.

We will now, as we proposed in § 1, split up formula (14) into two homogeneous linear substitutions by simply writing:

$$\text{(16)} \qquad \begin{cases} z'_1 = (d+ic)z_1 - (b-ia)z_2, \\ z'_2 = (b+ia)z_1 + (d-ic)z_2. \end{cases}$$

Here a, b, c, d, denote, according to formula (12), in the first place, arbitrary *real* magnitudes, which are subject to the condition:

$$a^2+b^2+c^2+d^2=1.$$

Meanwhile we may remark that the same formula, with this condition maintained, provided we regard a, b, c, d as susceptible of arbitrary complex values, represents at the same time the most general binary linear substitution of determinant 1. Hereby the formulæ of composition, which we shall immediately establish, acquire a more general significance, which, however, in the developments to which we must here limit ourselves, need not be further considered.

To deduce the formulæ of composition in question, let

$$S \begin{cases} z'_1 = (d+ic)z_1 - (b-ia)z_2, \\ z'_2 = (b+ia)z_1 + (d-ic)z_2, \end{cases}$$

be a first substitution, and similarly

$$T \begin{cases} z''_1 = (d'+ic')z'_1 - (b'-ia')z'_2, \\ z''_2 = (b'+ia')z'_1 + (d'-ic')z'_2, \end{cases}$$

a second. *We obtain the substitution ST, arising from the com-*

position of these, by eliminating z'_1, z'_2, *from the two systems.* We naturally put the result again in the form (16), and so write:

$$ST\begin{cases} z''_1=(d''+ic'')z_1-(b''-ia'')z_2, \\ z''_2=(b''+ia'')z_1+(d''-ic'')z_2. \end{cases}$$

Then direct comparison gives the following simple result:

$$(17)\qquad \begin{cases} a''=(ad'+a'd)-(bc'-b'c), \\ b''=(bd'+b'd)-(ca'-c'a), \\ c''=(cd'+c'd)-(ab'-a'b), \\ d''=-aa'-bb'-cc'+dd'. \end{cases}$$

We have thus, as we may observe, the symbolic notation ST applied in the same sense as in the preceding chapter, if we effect first the substitution S, then the substitution T.

We shall immediately apply the formulæ (14), (16), (17), in the establishment of the groups of substitutions which now correspond to the groups of rotations of the preceding chapter. First, however, we must consider the significance which these formulæ claim in a more general sense. That it was proper, in the treatment of rotations round a fixed point, to introduce the parameters a, b, c, d, of the preceding paragraph $\left(\text{or at least their quotients } \frac{a}{d}, \frac{b}{d}, \frac{c}{d}\right)$, *Euler* had already found.* It appears, however, that the formulæ of composition (17) remained still unknown for a long time, till they were discovered by *Rodrigues*† (1840). *Hamilton* then made the same formulæ the foundation ‡ of his calculus of quaternions, without at first recognising their significance for the composition of rotations, which was soon brought to light by *Cayley*.§ But the relation of these formulæ to the composition of binary linear substitutions remained still unobserved; to *Herr Laguerre* is due the

* "Novæ Commentationes Petropolitanæ," t. 20, p. 217.

† "Journal de Liouville," 1 série, tome v: "Des lois géométriques qui régissent le déplacement," &c.

‡ In fact, if we consider the quaternions:

$$q=ai+bj+ck+d,\ q'=a'i+b'j+c'k+d',$$

the product thereof:

$$qq'=q''=a''i+b''j+c''k+d''$$

is exactly given by the formulæ (17) of the text. It is interesting to consult the first reports of Hamilton on his calculus of quaternions, especially his letter to Graves in the "Philosophical Magazine," 1844, ii, p. 489.

§ "Philosophical Magazine," 1843, i, p. 141.

credit of having first recognised this connection on the formal side.* It first acquired a real importance by *Riemann's* interpretation of $(x+iy)$ on the sphere, and especially by *Cayley's* † formula (14).

§ 4. Return to the Groups of Substitutions—The Cyclic and Dihedral Groups.

We now proceed to establish the homogeneous linear substitutions of determinant 1,‡ which correspond, in the sense of formulæ (14), (16), to the groups of rotations previously investigated. Of course, the substitutions which we in this manner obtain are, on account of the double sign of the parameters a, b, c, d, *double as numerous* as the rotations from which we start. The group of substitutions is, therefore, in the first place *hemihedrically* isomorphic with the group of rotations; the question whether we cannot so limit or modify the group of substitutions that *simple* isomorphism ensues, will not be investigated till a later paragraph.

As regards the general rules of which we shall make use in establishing the groups of substitutions, we shall of course, in each case, provide for the system of co-ordinates a position as simple as possible; and besides this, we shall recur to the propositions which we have established in §§ 12 and 13 of the preceding chapter, with reference to the generation of the several groups of rotations.

In the case of the *cyclic groups* and the *dihedral groups*, the matter is so simple that we can write down the formulæ without more ado. It seems most convenient to let the two *poles* considered in connection with these groups coincide with the points $z=0$ and $z=\infty$. Then we have, for the rotations of the cyclic groups:

* "Journal de l'École polytechnique," cah. 42 (1867): "Sur le calcul des systèmes linéares."

† *Cf.* especially, too, M. Stephanos' article, "Mémoire sur la représentation des homographies binaires par des points de l'espace avec application à l'étude des rotations sphériques," Math. Ann., Bd. xxii. (1883), and also his note "Sur la théorie des quaternions" (*ibid.*).

‡ Or, as I shall say in future for brevity, where there is no fear of misunderstanding: the "homogeneous substitutions."

$$a=b=0,\ c=\sin\frac{\alpha}{2},\ d=\cos\frac{\alpha}{2},\ \alpha=\frac{2k\pi}{n},$$

and therefore for the $2n$ *homogeneous substitutions of the cyclic group:*

$$(18)\quad z'_1=e^{\frac{ik\pi}{n}}\cdot z_1,\ z'_2=e^{\frac{-ik\pi}{n}}\cdot z_2.\quad (k=0,\ 1,\ \ldots\ (2n-1)).$$

If we now proceed to the dihedral group, we shall choose one of the secondary axes, so that it coincides with the ξ-axis of our co-ordinate system in space (and therefore joins the points $z=+1$ and $z=-1$ on the sphere). We find for the corresponding revolution:

$$(19)\qquad z'_1=\mp iz_2,\ z'_2=\mp iz_1,$$

and therefore, by combination with (18), *for the* $4n$ *homogeneous substitutions of the dihedral group:*

$$(20)\quad \begin{cases} z'_1=e^{\frac{ik\pi}{n}}\cdot z_1,\ z'_2=e^{\frac{-ik\pi}{n}}\cdot z_2; \\ z'_1=ie^{\frac{-ik\pi}{n}}\cdot z_2,\ z'_2=ie^{\frac{ikn}{\pi}}\cdot z_1. \end{cases}\quad (k=0,\ 1,\ \ldots\ (2n-1)).$$

Account has already been taken in these formulæ of the double sign of (19), since we have allowed k to range, not merely from 0 to $(n-1)$, but from 0 to $(2n-1)$.

We have in particular, as we will note expressly, for the *quadratic group* the following 8 homogeneous substitutions:

$$(21)\quad \begin{cases} z'_1=\quad i^k\cdot z_1,\ z'_2=(-i)^k\cdot z_2; \\ z'_1=-(-i)^k\cdot z_2,\ z'_2=\quad i^k\cdot z_1; \end{cases}$$
$$(k=0,\ 1,\ 2,\ 3).$$

§ 5. The Groups of the Tetrahedron and Octahedron.

In the case of the tetrahedron and octahedron we shall distinguish two different positions of the system of co-ordinates. In the first case we allow, as appears most natural, the 3 co-ordinate axes ξ, η, ζ, of our co-ordinate system in space to simply coincide with the diagonals of the octahedron. In the second case we rotate the co-ordinate system so obtained on its ζ-axis through 45°, viz., so that (as proves advantageous later on) the $\xi\zeta$-plane coincides with a plane of symmetry of the tetrahedron.

Let us begin with the consideration of the *former* position. We can then make immediate use of formulæ (21), just written down, for the representation of the quadratic group. Recalling now, with regard to the generation of the tetrahedral and octahedral groups, the data which we have prepared in § 13, we will first construct the homogeneous substitutions which correspond to the two rotations (U and U^2) of period 3 round one of the diagonals of the corresponding cube. Evidently 2 diametrically opposite summits of the cube have the coordinates:

$$\xi=\eta=\zeta=\pm\frac{1}{\sqrt{3}},$$

and since:

$$\cos\frac{\pi}{3}=\frac{1}{2}=-\cos\frac{2\pi}{3},\ \sin\frac{\pi}{3}=\frac{\sqrt{3}}{2}=\sin\frac{2\pi}{3},$$

we obtain for the homogeneous substitutions for which these two summits remain unmoved (neglecting the double sign, which occurs again here):

$$a=b=c=\pm d=\frac{1}{2}.$$

Corresponding to this we have the two substitutions:

$$z'_1=\frac{(\pm 1+i)z_1-(1-i)z_2}{2},\quad z'_2=\frac{(1+i)z_1+(\pm 1-i)z_2}{2}.$$

Combining these now in a proper manner with the substitutions (21), *we obtain, for the right sides of the* 24 *homogeneous tetrahedral substitutions, the following pairs of linear expressions:*

$$(22)\ \begin{cases} i^k\,.\,z_1, & (-i)^k\,.\,z_2\,; \\ -(-i)^k\,.\,z_2, & i^k\,.\,z_1\,; \\ i^k\,.\,\dfrac{(\pm 1+i)z_1-(1-i)z_2}{2}, & (-i)^k\,.\,\dfrac{(1+i)z_1+(\pm 1-i)z_2}{2} \\ -(-i)^k\,.\,\dfrac{(1+i)z_1+(\pm 1-i)z_2}{2}, & i^k\,.\,\dfrac{(\pm 1+i)z_1-(1-i)z_2}{2} \end{cases}$$

$$(k=0,\ 1,\ 2,\ 3).$$

We pass to the octahedral group by adding a rotation V through $\frac{\pi}{2}$ round one of the 3 co-ordinate axes, say the ζ-axis.

For one of the two corresponding homogeneous substitutions we have manifestly:

$$(23)\qquad z'_1=\frac{1+i}{\sqrt{2}}\cdot z_1,\ z'_2=\frac{1-i}{\sqrt{2}}\cdot z_2.$$

In correspondence with this, we obtain the right-hand sides of the 24 homogeneous octahedral substitutions still wanting in the table (22), by multiplying, in each case, the left-hand one of the 24 linear expressions included in this table by $\frac{1+i}{\sqrt{2}}$, *the right-hand one by* $\frac{1-i}{\sqrt{2}}$.

It will be unnecessary to write down specially the new expressions here.

With regard, now, to the *second* position of the co-ordinate system relatively to our configuration, it is sufficient, in order to have the substitution formulæ belonging to it, to take account—in the formulæ (22), (23), &c., just obtained—of the transformation of co-ordinates which leads from the first position to the second. For such a transformation of co-ordinates, the original $\frac{z_1}{z_2}$ will be replaced by $\frac{1+i}{\sqrt{2}}\cdot\frac{z_1}{z_2}$, and of course simultaneously the original $\frac{z'_1}{z'_2}$ by $\frac{1+i}{\sqrt{2}}\cdot\frac{z'_1}{z'_2}$.* Let us observe, moreover, that $\frac{1+i}{\sqrt{2}}\cdot\frac{1-i}{\sqrt{2}}=1$. We thus obtain on brief reflexion the rule:

If we desire substitution formulæ which correspond to the new position of the system of co-ordinates, we must, in the expressions occurring on the left-hand side in (22), *leave* z_1 *unaltered, and replace* z_2 by $\frac{1-i}{\sqrt{2}}\cdot z_2$; *on the other hand, in the expressions occurring there on the right-hand side, we must replace* z_1 *by* $\frac{1+i}{\sqrt{2}}\cdot z_1$, *and leave* z_2 *unaltered.*

With such entirely elementary operations I again omit to explicitly note the expressions which occur.

* Namely, if we suppose the rotation through 90° round $O\zeta$-axis proceed in a positive sense.

§ 6. The Icosahedral Group.

We have now to investigate the homogeneous substitutions of the icosahedron. With this object we will assign to the icosahedron such a position with regard to the system of co-ordinates that the rotation through $\frac{2\pi}{5}$, which we previously (§ 12 of the preceding chapter) denoted by S, takes place in a positive sense round the ζ-axis, while at the same time the cross-line, round which the revolution U (*loco cito*) takes place, coincides with the η-axis. *Then we have at once, corresponding to the operations S, U, the following substitutions:*

$$(24)\qquad \begin{cases} S: \begin{cases} z'_1 = \pm\epsilon^3 z_1, \\ z'_2 = \pm\epsilon^2 z_2; \end{cases} \\ U: \begin{cases} z'_1 = \mp\ z_2, \\ z'_2 = \pm\ z_1, \end{cases} \end{cases}$$

which, taken together, generate the dihedral group belonging to the vertical diagonal of the icosahedron.* By ϵ here, as always for the future, the fifth root of unity:

$$(25)\qquad \epsilon = e^{\frac{2i\pi}{5}}$$

is to be understood.

Our convention respecting the position of the system of co-ordinates admits of a twofold possibility with regard to the revolution T, which we have now still to consider.

The axis of T can move in the $\xi\zeta$-plane either through the first and third quadrants of the system of co-ordinates $\xi\zeta$, or through the second and fourth. *We will agree that the latter is the case.* If we understand by γ the acute angle which the said axis makes with $O\zeta$, one of its ends will have the co-ordinates:

$$\xi = -\sin\gamma,\ \eta = 0,\ \zeta = \cos\gamma,$$

and the parameters of the corresponding rotation become by (12) (since we are concerned with a rotation through 180°):

$$a = \mp\sin\gamma,\ b = 0,\ c = \pm\cos\gamma,\ d = 0;$$

* This is here related to a somewhat different system of co-ordinates to that of formula (20).

where, as always in these formulæ, the upper and the lower signs go together respectively.

The question now is how we calculate the angle γ. For this purpose I will return to the parameters of S (24):

$$a'=b'=0,\ c'=\pm\sin\frac{\pi}{5},\ d'=\pm\cos\frac{\pi}{5};$$

and to the formulæ of composition (17). By means of this formula we find for the parameter d'' of the operation ST:

$$\begin{aligned} d'' &= -aa'-bb'-cc'+dd' \\ &= \pm\cos\gamma\cdot\sin\frac{\pi}{5}. \end{aligned}$$

Now the operation ST (as a glance at the figure of the icosahedron shows) has the period 3; it must therefore be identical with $\pm\cos\frac{\pi}{3}=\pm\frac{1}{2}$. We thus obtain, if we further consider that $\cos\gamma$ must be positive:

$$\cos\gamma\,.\,\sin\frac{\pi}{5}=\frac{1}{2},$$

or, if we again introduce the root of unity ϵ, and recall that

$$(\epsilon^2-\epsilon^3)(\epsilon^4-\epsilon)=\epsilon+\epsilon^4-\epsilon^2-\epsilon^3=\sqrt{5},$$

we must have:

$$\cos\gamma=\frac{\epsilon-\epsilon^4}{i\sqrt{5}},$$

and, therefore, again assuming the positive sign,

$$\sin\gamma=\frac{\epsilon^2-\epsilon^3}{i\sqrt{5}},$$

We now introduce these values into the expressions a, b, c, d, just given, and also refer to the formulæ (16). *Then we have, finally, for the two homogeneous substitutions which correspond to the rotation* T:

$$(26)\qquad T:\begin{cases}\sqrt{5}\,.\,z'_1=\mp(\epsilon-\epsilon^4)z_1\pm(\epsilon^2-\epsilon^3)z_2,\\ \sqrt{5}\,.\,z'_2=\pm(\epsilon^2-\epsilon^3)z_1\pm(\epsilon-\epsilon^4)z_2.\end{cases}$$

From (24), (26), we now construct at once the whole set of icosahedral substitutions. We need only remember that we previously brought the icosahedral rotations into the following table:

$$S^\mu,\ S^\mu U,\ S^\mu T S^\nu,\ S^\mu T S^\nu U,\ (\mu,\ \nu=0,\ 1,\ 2,\ 3,\ 4).$$

Corresponding to this we obtain for the 120 *homogeneous icosahedral substitutions:*

$$(27)\begin{cases} S^\mu: \begin{cases} z'_1=\pm\epsilon^{3\mu}\,.\,z_1,\\ z'_2=\pm\epsilon^{2\mu}\,.\,z_2; \end{cases}\\ S^\mu U: \begin{cases} z'_1=\mp\epsilon^{2\mu}\,.\,z_2,\\ z'_2=\pm\epsilon^{3\mu}\,.\,z_1; \end{cases}\\ S^\mu T S^\nu: \begin{cases} \sqrt{5}\,.\,z'_1=\pm\epsilon^{3\nu}\left(-(\epsilon-\epsilon^4)\,\epsilon^{3\mu}\,.\,z_1+(\epsilon^2-\epsilon^3)\,\epsilon^{2\mu}\,.\,z_2\right),\\ \sqrt{5}\,.\,z'_2=\pm\epsilon^{2\nu}\left(+(\epsilon^2-\epsilon^3)\,\epsilon^{3\mu}\,.\,z_1+(\epsilon-\epsilon^4)\,\epsilon^{2\mu}\,.\,z_2\right); \end{cases}\\ S^\mu T S^\nu U: \begin{cases} \sqrt{5}\,.\,z'_1=\mp\epsilon^{2\nu}\left(+(\epsilon^2-\epsilon^3)\,\epsilon^{3\mu}\,.\,z_1+(\epsilon-\epsilon^4)\,\epsilon^{2\mu}\,.\,z_2\right),\\ \sqrt{5}\,.\,z'_2=\pm\epsilon^{3\nu}\left(-(\epsilon-\epsilon^4)\,\epsilon^{3\mu}\,.\,z_1+(\epsilon^2-\epsilon^3)\,\epsilon^{2\mu}\,.\,z_2\right). \end{cases} \end{cases}$$

I will further call attention to the simple rule by which here (as also in the previous cases), the periodicity of the individual rotation is determined by formula (15). We obtain, in virtue of this formula, for the angle a of a rotation $S^\mu T S^\nu$:

$$\cos\frac{a}{2}=\mp\frac{(\epsilon-\epsilon^4)\ (\epsilon^{3\mu+3\nu}-\epsilon^{2\mu+2\nu})}{2\sqrt{5}},$$

and analogously for the angle of rotation of $S^\mu T S^\nu U$:

$$\cos\frac{a}{2}=\mp\frac{(\epsilon^2-\epsilon^3)\ (\epsilon^{3\mu+2\nu}-\epsilon^{2\mu+3\nu})}{2\sqrt{5}}.$$

We have, therefore, for $S^\mu T S^\nu$ *the period* 2, *if* $\mu+\nu\epsilon\equiv0$, *for* $S^\mu T S^\nu U$ *if* $3\mu+2\nu\equiv0$ (*mod.* 5).

We have for $S^\mu T S^\nu$ *the period* 3, *if* $\mu+\nu\equiv\pm1$, *for* $S^\mu T S^\nu U$, *if* $3\mu+2\nu\equiv1\pm$ (*mod.* 5).

In the 20 *other cases* $S^\mu T S^\nu$ *and* $S^\mu T S^\nu U$ *are respectively of period* 5. To this must be added, what is self-evident, that all the $S^\mu U$ have the period 2; all the S^μ, with the sole exception of S° (identity), have the period 5.

§ 7. Non-Homogeneous Substitutions—Consideration of the Extended Groups.

From the homogeneous substitutions we naturally descend without calculation to the non-homogeneous substitutions. I exhibit, however, the formulæ in question in a tabular collection, because, when the fixed value of the substitution-determinant hitherto maintained is abolished, they admit of a certain amount of condensation, and hence, in fact, become very readily surveyed. *We find for the non-homogeneous substitutions:*

(i.) For the *cyclic group:*

(28) $$z' = e^{\frac{2ik\pi}{n}} \cdot z, \left(k=0, 1, \ldots (n-1)\right);$$

(ii.) For the *dihedron:*

(29) $$z' = e^{\frac{2ik\pi}{n}} \cdot z, \quad z' = -\frac{e^{-\frac{2ik\pi}{n}}}{z}, \text{ (}k\text{ as before)};$$

(iii.) For the *tetrahedron* and first assumption respecting the position of the system of co-ordinates:

(30a) $$z' = \pm z, \ \pm\frac{1}{z}, \ \pm i \cdot \frac{z+1}{z-1}, \ \pm i \cdot \frac{z-1}{z+1}, \ \pm\frac{z+i}{z-i}, \ \pm\frac{z-i}{z+i},$$

also, for the other assumption:

(30b) $$z' = \pm z, \ \pm\frac{i}{z}, \ \pm\frac{(1+i)z+\sqrt{2}}{\sqrt{2}\cdot z-(1-i)}, \ \pm\frac{\sqrt{2}\cdot z-(1-i)}{(1+i)z+\sqrt{2}},$$
$$\pm\frac{(1-i)z+\sqrt{2}}{\sqrt{2}\cdot z-(1+i)}, \ \pm\frac{\sqrt{2}\cdot z-(1+i)}{(1-i)z+\sqrt{2}};$$

(iv.) For the *octahedron* with similar distinctions in the two cases:

(31a) $$z' = i^k z, \ \frac{i^k}{z}, \ i^k \cdot \frac{z+1}{z-1}, \ i^k \cdot \frac{z-1}{z+1}, \ i^k \cdot \frac{z+i}{z-i}, \ i^k \cdot \frac{z-i}{z+i},$$

and:

(31b) $$z' = i^k \cdot z, \ \frac{i^k}{z}, \ i^k \cdot \frac{(1+i)z+\sqrt{2}}{\sqrt{2}\cdot z-(1-i)}, \ i^k \cdot \frac{\sqrt{2}\cdot z-(1-i)}{(1+i)z+\sqrt{2}},$$
$$i^k \cdot \frac{(1-i)z+\sqrt{2}}{\sqrt{2}\cdot z-(1+i)}, \ i^k \cdot \frac{\sqrt{2}\cdot z-(1+i)}{(1-i)z+\sqrt{2}};$$

k has here in each case to assume in succession the values 0, 1, 2, 3.

(v.) For the *icosahedron:*

$$(32)\qquad z'=\epsilon^{\mu}z,\ \frac{-\epsilon^{4\mu}}{z},\ \epsilon^{\nu}\cdot\frac{-(\epsilon-\epsilon^4)\epsilon^{\mu}\cdot z+(\epsilon^2-\epsilon^3)}{(\epsilon^2-\epsilon^3)\epsilon^{\mu}\cdot z+(\epsilon-\epsilon^4)},$$

$$-\epsilon^{4\nu}\cdot\frac{(\epsilon^2-\epsilon^3)\epsilon^{\mu}\cdot z+(\epsilon-\epsilon^4)}{-(\epsilon-\epsilon^4)\epsilon^{\mu}\cdot z+(\epsilon^2-\epsilon^3)},$$

$$\left(\epsilon=e^{\frac{2i\pi}{5}};\ \mu,\ \nu,=0,\ 1,\ 2,\ 3,\ 4\right).$$

From these formulæ we now pass at once to those which correspond to *extended groups* (as we expressed it in Chapter I.), namely, if we deduct the single groups of formulæ (30a), the $\xi\zeta$-plane is throughout a plane of symmetry for the configuration just considered. Now we can generate the extended group by combining the reflexion on this very plane of symmetry with the rotations of the original group. This reflexion is, however, given analytically by the formula:

$$(33)\qquad z'=\bar{z},$$

where $\bar{z}$ denotes the conjugate value of the imaginary quantity z. *Hence we shall obtain formulæ for the operations of the extended group if we place alongside of the formulæ* (28) *to* (32) ((30a) *alone excepted*) *the others in which z is replaced by $\bar{z}$.*

I conclude this paragraph with two short historical remarks. Of the groups of substitutions (28) to (32) only two cases come particularly into prominence in earlier literature (except the cyclic groups, which, of course, occur everywhere), viz., the dihedral group for $n=3$ and the octahedral group (31a). The first case appears in a form somewhat different to that of (29), but only because a different system of co-ordinates is established on the z-sphere, viz., that for which that great circle which we have hitherto described as the equator coincides with the meridian of real numbers, and the summits of the dihedron have the arguments $z=0,\ 1,\ \infty$. We thus find the formulæ:

$$z'=z,\ \frac{1}{z},\ 1-z,\ \frac{1}{1-z},\ \frac{z}{z-1},\ \frac{z-1}{z},$$

which in projective geometry connect the 6 corresponding values of the double ratio and in the theory of elliptic functions (which is really the same thing) the 6 corresponding values of k^2 (the square of Legendre's modulus). The group (31a) is found in several places in Abel's works.* The object is there to present the different values of k^2, which result on transforming a given elliptic integral of the first kind by a *linear* substitution into Legendre's normal form:

$$\int\frac{dx}{\sqrt{1-x^2 \,.\, 1-k^2x^2}}.$$

Abel remarks that these different values are represented in terms of any one of them in the following manner:

$$k^2,\ \frac{1}{k^2},\ \left(\frac{1+\sqrt{k}}{1-\sqrt{k}}\right)^4,\ \left(\frac{1-\sqrt{k}}{1+\sqrt{k}}\right)^4,\ \left(\frac{i+\sqrt{k}}{i-\sqrt{k}}\right)^4,\ \left(\frac{i-\sqrt{k}}{i+\sqrt{k}}\right)^4.$$

If we here extract the fourth root and replace $\sqrt{k}$ by z all through, these are evidently exactly the expressions (31a).

§ 8. Simple Isomorphism in the Case of Homogeneous Groups of Substitutions.

For a discussion of the groups of substitutions now obtained from the point of view of the theory of groups, it will be sufficient to refer here to the analogous inquiries in our first chapter. In fact, our non-homogeneous groups of substitutions are simply isomorphic with the groups of rotations there considered, the homogeneous ones at least hemihedrically, where let us expressly remark, that among the homogeneous substitutions the two:

$$\left.\begin{matrix} z'_1 = z_1 \\ z'_2 = z_2 \end{matrix}\right\} \text{ and } \begin{matrix} z'_1 = -z_1 \\ z'_2 = -z_2 \end{matrix}$$

always correspond to "identity."

Moreover, we will concern ourselves with a question of an allied nature, certainly, if not purely one belonging to the theory of groups, a question which we have already pointed out (§ 4 *supra*), and the answering of which will be of prime importance to us in the sequel. We have found for a group of N rotations in every case $2N$ homogeneous substitutions. We ask if it be not possible to extract from among these $2N$ substitutions N of them forming a group so that *simple*

* See, *e.g.*, Bd. i, p. 259 (new edition by Sylow and Lie).

isomorphism with the group of rotations ensues, or if we cannot at least attain that isomorphism, by imparting any other value to the determinant, which we have hitherto taken $+1$, of the individual substitutions?

We begin with the repetitions of a single rotation, *i.e.*, with the *cyclic groups*, where, in order not to apparently limit the investigation by the introduction of a canonical system of co-ordinates, we will start from a perfectly arbitrary system of co-ordinates. We therefore take, say, a rotation through $\frac{2\pi}{n}$, for which an arbitrary point ξ, η, ζ, on our sphere remains unmoved. To the corresponding linear substitution (16):

$$z'_1=(d+ic)z_1-(b-ia)z_2,$$
$$z'_2=(b+ia)z_1+(d-ic)z_2,$$

we have hitherto attached the parameters:

$$a=\pm\xi\sin\frac{\pi}{n},\ b=\pm\eta\sin\frac{\pi}{n},\ c=\pm\zeta\sin\frac{\pi}{n},\ d=\pm\cos\frac{\pi}{n}.$$

We will now write instead of them, *taking the determinant of the substitution equal to* ρ^2:

$$(34)\qquad a_1=\rho\xi\sin\frac{\pi}{n},\ b_1=\rho\eta\sin\frac{\pi}{n},\ c_1=\rho\zeta\sin\frac{\pi}{n},\ d_1=\rho\cos\frac{\pi}{n}.$$

Recurring then to the formulæ of composition (17), we obtain for the parameters of the kth repetition of our substitution:

$$a_k=\rho^k\,.\,\xi\sin\frac{k\pi}{n},\ b_k=\rho^k\,.\,\eta\sin\frac{k\pi}{n},\ c_k=\rho^k\,.\,\zeta\sin\frac{k\pi}{n},\ d_k=\rho^k\cos\frac{k\pi}{n}.$$

We require now—in order that simple isomorphism with the corresponding group of rotations may take place—that the nth repetition of our substitution should be identity, and that, therefore:

$$a_n=b_n=c_n=0,\ d_n=1.$$

It is clearly necessary for this that:

$$\rho^n=-1.$$

We shall, therefore, then, and only then, attain to simple isomorphism between the substitutions and the group of rotations when we introduce in (34) ρ *as the* n^{th} *root of* (-1). Hereby, however, the value ρ^2 of the determinant of the substitutions is determined, or at least limited to a few possibilities only. If n is odd, we can take $\rho=-1$, and therefore the determinant $=+1$. If n is even, the value $+1$ of the determinant of the substitution is inevitable. In particular, if $n=2$, we must choose the determinant $= -1$, and the magnitude $\rho = \pm i$.

We now consider the *dihedral* group. We have for it, first of all, the rotations S^μ (with $S^n=1$), which, according to what has just been said, we must make correspond to substitutions of determinant $\rho^{2\mu}$, where $\rho^n = -1$. We have further the rotations $S^\mu T$ of period 2. To effect simple isomorphism we shall certainly provide the substitution which corresponds to T with the determinant (-1). Now we know that in the compositions of two substitutions their determinants are *multiplied*. Therefore we obtain for $S^\mu T$ a substitution of determinant $-\rho^{2\mu}$. But this must itself again be equal to -1, because $S^\mu T$ has the period 2. Thus we have for ρ the simultaneous equations:

$$\rho^n = -1,\ \rho^{2\mu} = +1,\ \big(\mu = 0,\ 1,\ .\ .\ .\ (n-1)\big).$$

These are evidently only reconcilable when n is odd (whence $\rho = -1$). Therefore it follows that, *in the case of the dihedral group, the desired simple isomorphism can only exist for n odd, never for n even.*

We shall in the sequel lay special stress on the negative part of this proposition, for we at once deduce from it an analogous theorem for the groups of the tetrahedron, octahedron, and icosahedron. *In the case of the tetrahedron, octahedron, and icosahedron, simple isomorphism between the group of rotations and the group of homogeneous substitutions is impossible.* They all contain, namely, as sub-group at least one dihedral group with n even (viz., a quadratic group), and herein, as we have just seen, lies the impossibility alluded to.

§ 9. Invariant Forms belonging to a Group—The Set of Forms for the Cyclic and Dihedral Groups.

True to the general process of thought which we have sketched in § 1 of this chapter, we now ask—after finding the homogeneous group of substitutions which correspond to the several groups of rotations—for all such *forms* $F(z_1, z_2)$ as remain unaltered, save as to a factor, for these substitutions. Such an *invariant form* (an expression which we shall hereafter retain) clearly represents, when equated to zero, a system of points on our sphere which remain unaltered for all rotations of the group in question—a proposition which we can reverse. Now such a system of points must necessarily separate into mere groups of points of the kind which we have described in § 10 of the preceding chapter as *appertaining to the group*. The invariant forms which we seek therefore arise when any number of the forms which correspond to the aforesaid groups of points are multiplied together.

Concerning the nature of the *ground-forms* thus presenting themselves, we can *a priori* make certain more detailed statements. If N is the number of rotations of a group, the groups of points which appertain to them consist in general of N separate points. The general ground-form will accordingly be a form of the Nth degree, and will contain besides—corresponding to the singly infinite number of groups of points mentioned before—an essential (not merely factorial) parameter. But there occur among the general groups of points those in particular which contain only a smaller number of separate points. In accordance with this, *special ground-forms*, of degree $\frac{N}{\nu}$, will occur, which can only be considered as a special case of the general ground-form when we raise them to the νth power.

If we wish to push these general results any further, we must separate here the case of the cyclic groups from the others.

In the case of the *cyclic groups* there occur among the general groups of points only two special ones, each consisting of only one point, viz., one of the two poles. *Accordingly in their case there are two special ground-forms, and these linear ones.* Retaining the system of co-ordinates which was introduced in § 4 in the treatment of the cyclic groups, these are simply z_1 and z_2 themselves. But further, we can here very easily construct the general ground-forms, and this by means of a method of reasoning which we shall find exceedingly useful in the following cases. To pass to the general ground-forms we construct the nth powers of z_1 and z_2, and convince ourselves that, by the several substitutions (18), they acquire the factor $(-1)^k$. Whence we conclude that $\lambda_1 z_1{}^n + \lambda_2 z_2{}^n$, understanding by $\lambda_1 : \lambda_2$ an arbitrary parameter, is also an invariant form in each case. Since its degree is equal to n (equal to the number of rotations of the group), it is at the same time a ground-form. *It is manifestly, without further proof, the general ground-form.* For we can so determine $\lambda_1 : \lambda_2$ that $\lambda_1 z_1^n + \lambda_2 z_2^n$ vanishes for an arbitrary point on the sphere, and therefore just represents the group of points proceeding from it by means of the rotations of the cyclic groups. Thus we have given a general solution, for the case of cyclic groups, of the questions which first confronted us. We can express the result by saying that *for the cyclic groups* (18) *the most general invariant form is given by:*

$$(35) \qquad z_1^{\alpha} \cdot z_2^{\beta} \cdot \prod_i (\lambda_1^{(i)} z_1^n + \lambda_2^{(i)} z_2^n),$$

where α, β, *denote any positive integral numbers and* $\lambda_1^{(i)}$, $\lambda_2^{(i)}$, *any parameters.*

In the other cases the theory presents certain differences, but only in so far as for them, among the general groups of N separate points each, *three groups of a smaller number of points* occur. For the multiplicities which are to be attributed to these special cases, so far as we include them under the general groups of points, we will again assume the notation ν_1, ν_2, ν_3, which we used in § 9 of the preceding chapter. The said groups of points then contain respectively $\frac{N}{\nu_1}$, $\frac{N}{\nu_2}$, $\frac{N}{\nu_3}$, separate points, and produce accordingly 3 special ground-forms F_1, F_2, F_3, respectively of the same degree. We construct $F_1^{\nu_1}$, $F_2^{\nu_2}$, $F_3^{\nu_3}$. Then it is shown that these powers all assume *the same* constant factor for the homogeneous substitutions in each case concerned. *Therefore every linear combination:*

$$\lambda_1 F_1^{\nu_1} + \lambda_2 F_2^{\nu_2} + \lambda_3 F_3^{\nu_3},$$

is an invariant form, and, indeed, as its degree shows, a ground-form.

But the general ground-form contains, as we have said, only one essential parameter, while we here have two in $\lambda_1 : \lambda_2 : \lambda_3$. We conclude that for the representation of all ground-forms it suffices to take into consideration the linear combinations:

$$\lambda_1 F_1^{\nu_1} + \lambda_2 F_2^{\nu_2},$$

and *that therefore an identity:*

$$(36)\qquad \lambda_1^{(0)}F_1^{\nu_1}+\lambda_2^{(0)}F_2^{\nu_2}+\lambda_3^{(0)}F_3^{\nu_3}=0,$$

must exist between F_1, F_2, F_3.

Considering $F_3^{\nu_3}$ always eliminated by means of this identity, we have finally, as the expression of the most general invariant form:

$$(37)\qquad F_1^{\alpha}\,.\,F_2^{\beta}\,.\,F_3^{\gamma}\,.\prod_i(\lambda_1^{(i)}F_1^{\nu_1}+\lambda_2^{(i)}F_2^{\nu_2}),$$

where the positive integral numbers α, β, γ, and the parameters $\lambda_1^{(i)}$, $\lambda_2^{(i)}$, are throughout arbitrary.

In the case of the *dihedron*, the whole theory here described presents itself again in such a simple form, in virtue of the position of the system of co-ordinates established in § 4, that we can write down the result immediately. We have:

$$N=2n,\ \nu_1=\nu_2=2,\ \nu_3=n,$$

and find accordingly:

$$(38)\qquad F_1=\frac{z_1^n+z_2^n}{2},\quad F_2=\frac{z_1^n-z_2^n}{2},\quad F_3=z_1z_2.$$

$F_2=0$ represents the summits of the dihedron. $F_1=0$ the mid-edge points, $F_3=0$ the pair of poles. Between F_1, F_2, F_3, exists then in correspondence with (36) the identity:

$$(39)\qquad F_1^2-F_2^2-F_3^n=0.$$

As regards the *tetrahedron*, *octahedron*, and *icosahedron*, the establishment of the special ground-forms requires in their case special considerations, to which we now turn.*

* The forms F_1, F_2, F_3, considered in the several cases together with the relations subsisting between them, occur for the first time in Herr Schwarz's memoir: "Ueber diejenigen Fälle, in denen die Gaussische Reihe $F(\alpha,\beta,\gamma,x)$ eine algebraische Function ihres vierten Elementes ist," Borchardt's Journal, Bd. 75 (1872). See, too, frequent contributions in the Züricher Vierteljahrschrift from 1871 onwards. The reason of my only cursorily citing this fundamental work is that its point of view in the treatment of forms F is, in the first place, quite different from ours. Its starting-point is formed by certain questions in the theory of *the conformable representation*, on which we shall enter more fully in the following chapter. On the other hand, Herr Schwarz gives neither the groups of linear substitutions, nor the relation to the theory of invariants which we shall now lay so much stress on.

§ 10. Preparation for the Tetrahedral and Octahedral Forms.

In the case of the tetrahedron and octahedron we have to distinguish, in accordance with § 5, two positions of the system of co-ordinates. Beginning with the first of these, we find for the summits of the octahedron (*i.e.*, now the points of intersection of the co-ordinate axes with the sphere) the arguments:

$$z = 0, \ \infty, \ \pm 1, \ \pm i,$$

and *therefore the octahedron is simply given by the following equation:*

$$z_1 z_2 (z_1^4 - z_2^4) = 0. \tag{40}$$

In a similar manner we determine the equations for the two corresponding tetrahedra and the cube determined by its 8 summits. The 8 summits of the cube have as co-ordinates:

$$\pm\xi = \pm\eta = \pm\zeta = \frac{1}{\sqrt{3}}.$$

We shall pick out the summits of one of the corresponding tetrahedra, if we choose here, among the 8 possible combinations of sign, those 4 for which the product $\xi\eta\zeta$ is positive. Substituting in the formulæ (6), we obtain for the arguments of the 4 summits of the tetrahedron:

$$z = \frac{1+i}{\sqrt{3}-1}, \ \frac{1-i}{\sqrt{3}+1}, \ \frac{-1+i}{\sqrt{3}+1}, \ \frac{-1-i}{\sqrt{3}-1}.$$

Whence we obtain (by multiplying out the linear factors) the equation of the first tetrahedron in the form:

$$z_1^4 + 2\sqrt{-3} \,.\, z_1^2 z_2^2 + z_2^4 = 0. \tag{41}$$

In the same way we find for the counter-tetrahedron:

$$z_1^4 - 2\sqrt{-3} \,.\, z_1^2 z_2^2 + z_2^4 = 0, \tag{42}$$

and finally for the *cube*, on multiplying together the left sides of (41) and (42):

$$z_1^8 + 14 z_1^4 z_2^4 + z_2^8 = 0. \tag{43}$$

I will denote in the sequel the left sides of (40), (41), (42), (43), by t, Φ, Ψ, W. If we now rotate the system of co-ordinates, as we proposed at the end of § 5, through an angle of 45° round the ζ-axis, these forms are transformed into others with only real co-efficients. I shall distinguish these forms by accents, and put:

$$(44)\qquad \begin{cases} t' = z_1 z_2(z_1^4 + z_2^4), \\ \Phi' = z_1^4 + 2\sqrt{3}\,.\,z_1^2 z_2^2 - z_2^4, \\ \Psi' = z_1^4 + 2\sqrt{3}\,.\,z_1^2 z_2^2 - z_2^4, \\ W' = z_1^8 - 14 z_1^4 z_2^4 + z_2^8. \end{cases}$$

Equated to zero, these forms represent of course the octahedron, tetrahedron, and counter-tetrahedron, as well as the cube relatively to the new system of co-ordinates.

§ 11. The Set of Forms for the Tetrahedron.

In accordance with the explanations given in § 9, our whole consideration of the tetrahedral forms may now be limited to two points; first, to determine the constant factors to which the ground-forms:

$$(45)\qquad \begin{cases} \Phi = z_1^4 + 2\sqrt{-3}\,.\,z_1^2 z_2^2 + z_2^4, \\ \Psi = z_1^4 - 2\sqrt{-3}\,.\,z_1^2 z_2^2 + z_2^4, \\ t = z_1 z_2(z_1^4 - z_2^4), \end{cases}$$

or the corresponding ones Φ', Ψ', t' (44), are subject for the homogeneous substitutions of the tetrahedron; secondly, to note the linear identity which connects Φ^3, Ψ^3, t^2, or Φ'^3, Ψ'^3, t'^2, with one another.

With regard to the first, we recall the *generation* of the group of the tetrahedron as we established it in § 13 of the preceding chapter, and have already used it in the present chapter. For the substitutions of the quadratic group (21), Φ, Ψ, t, evidently remain in general unaltered. On the other hand, for those substitutions which correspond to the rotation U of period 3, Φ and Ψ receive factors $e^{\frac{2i\pi}{3}}$ and $e^{\frac{4i\pi}{3}}$, while t remains invariant for these also. The consequence is that, in addition to Φ^3 and Ψ^3, $\Phi\Psi = W$ also remains unaltered throughout, while Φ and Ψ themselves are only transformed into themselves by the substitutions of the quadratic group. As regards this latter circumstance, we perceive in it a confirmation of a principle which we can establish *a priori*. This asserts that *those substitutions of a homogeneous group, which leave altogether unaltered a corresponding invariant form, must form a self-conjugate sub-group within the main group of substitutions.* These remarks, of course, just apply to the forms Φ', ψ', t', W'.

Having confirmed by these remarks the existence of the supposed identity between Φ^3, Ψ^3, t^2, &c.,* we shall be able to compute it by only taking into consideration the first terms in the expressions of Φ^2, Ψ^3, t^2. In this way we find without trouble:

(46a) $$12\sqrt{-3}\,.\,t^2 - \Phi^3 + \Psi^3 = 0,$$

or:

(46b) $$12\sqrt{3}\,.\,t'^2 - \Phi'^3 + \Psi^{3'} = 0.$$

In connection with the results here obtained two remarks may be made which are both related to the invariant theory of binary forms, and of which the one may express the significance which the said theory will often have for us in the sequel, while the other is designed to marshal the results obtained by us in the case of the tetrahedron relatively to the otherwise well-known products of the invariant theory.

Suppose that, of the forms (45), we have only so far computed one, viz., Φ; then the theory of invariants supplies us with the means of deriving from it other tetrahedral forms by mere processes of differentiation. We have only to establish any *covariants* of Φ. In fact, if Φ is transformed into itself, save as to a factor, by any homogeneous linear substitutions, so also is every covariant; this is an immediate deduction from the definition of covariant forms. Now Φ is a binary form of the 4th order, and the theory of invariants shows† that such a form only possesses two independent covariants —the *Hessian* form of Φ and the *functional determinant of this form with* Φ. The former is of the 4th, the latter of the 6th degree; moreover, we may convince ourselves that the former is not identical with Φ. We, therefore, conclude at

* Since Φ^3, Ψ^3, t^2 remain uniformly unaltered by the tetrahedral substitutions (22).

† *Cf. e.g.*, Clebsch, "Theorie der binären algebraischen Formen" (Leipzig, 1872), p. 134, &c., or the other text-books of the theory of invariants, *e.g.*, Salmon-Fiedler, "Algebra der linearen Transformationen" (Leipzig, 2nd edition, 1877), Faà de Bruno-Walter, "Einleitung in die Theorie der binären Formen" (Leipzig, 1881), &c.

once that *the Hessian of Φ, equated to zero, represents the counter-tetrahedron, and similarly that the functional determinant, equated to zero, represents the corresponding octahedral form.* For both these forms, equated to zero, must represent such groups of points as remain unaltered by the tetrahedral rotations, and no other groups of only 4 or only 6 connected points can exist besides those just mentioned, or at least do not come under consideration (inasmuch as the 4 summits of the original tetrahedron, which likewise form such a group, are already given by $\Phi=0$). *We should therefore be able to calculate also amongst the forms* (45) *both Ψ and t by constructing the Hessian form of Φ, and then, from this and Φ, the functional determinant.* In fact, we get by calculating out directly:

$$\begin{vmatrix} \frac{\delta^2\Phi}{\delta z_1^2} & \frac{\delta^2\Phi}{\delta z_1 \delta z_2} \\ \frac{\delta^2\Phi}{\delta z_2 \delta z_1} & \frac{\delta^2\Phi}{\delta z_2^2} \end{vmatrix} = 48\sqrt{-3}\,.\,\Psi,$$

and:

$$\begin{vmatrix} \frac{\delta\Phi}{\delta z_1} & \frac{\delta\Phi}{\delta z_2} \\ \frac{\delta\Psi}{\delta z_1} & \frac{\delta\Psi}{\delta z_2} \end{vmatrix} = 32\sqrt{-3}\,.\,t.$$

The theory of invariants possesses, as we see, in virtue of these remarks, the character of a *method of computation.* As regards our further elaboration by the theory of binary invariants, let us recur to the general theory of biquadratic forms, let:

(47) $$F = a_0 z_1^4 + 4a_1 z_1^3 z_2 + 6a_2 z_1^2 z_2^2 + 4a_3 z_1 z_2^3 + a_4 z_2^4$$

be such a form. Then we have in the first place, as already explained, two covariants, which we will now denote by H and T, the numerical factors being properly determined:

(48) $$\left\{ H = \frac{1}{144}\cdot \begin{vmatrix} \frac{\delta^2 F}{\delta z_1^2} & \frac{\delta^2 F}{\delta z_1 \delta z_2} \\ \frac{\delta^2 F}{\delta z_2 \delta z_1} & \frac{\delta^2 F}{\delta z_2^2} \end{vmatrix}, \quad T = \frac{1}{8}\cdot \begin{vmatrix} \frac{\delta F}{\delta z_1} & \frac{\delta F}{\delta z_2} \\ \frac{\delta H}{\delta z_1} & \frac{\delta H}{\delta z_2} \end{vmatrix}. \right.$$

We have, further, two invariants:

(49) $$\left\{ g_2 = a_0 a_4 - 4a_1 a_3 + 3a_2^2, \quad g_3 = \begin{matrix} a_0 & a_1 & a_2 \\ a_1 & a_2 & a_3 \\ a_2 & a_3 & a_4 \end{matrix} \right|$$

(where I have applied on the left-hand side that notation to which I shall hereafter have anyhow to return in connection with Weierstrass's theory of elliptic functions). We have, finally, as the single relation between these forms, the following:

$$(50) \qquad 4H^3 - g_2HF^2 + g_3F^3 + T^2 = 0.$$

Let us now put our Φ in the place of F; then we have in the first place:

$$g_2 = 0.$$

This means, if we adopt the geometrical mode of expression which, *e.g.*, is explained by Clebsch, l. c. p. 171:

*The form Φ equated to zero represents an equianharmonic group of points.**

We find, further, for our Φ:

$$H = \frac{1}{\sqrt{-3}} \cdot \Psi, \; T = 4t, \; g_3 = \frac{-4}{3\sqrt{-3}}.$$

Hence the identity (46a) is included in the general relation (50) as a particular case, as was to be expected. We must, therefore, say that our geometrical reflexions on the group-theory have led us in the case of the tetrahedral forms not so much to new algebraical results, as to a new way to results otherwise known.

§ 12. The Set of Forms for the Octahedron.

Turning now to the octahedral forms, we already know, of the 3 special ground-forms appertaining to them, the two:

$$(51a) \qquad \begin{cases} t = z_1z_2(z_1^4 - z_2^4), \\ W = z_1^8 + 14\, z_1^4z_2^4 + z_2^8\,; \end{cases}$$

and:

$$(51b) \qquad \begin{cases} t' = z_1z_2(z_1^4 + z_2^4), \\ W' = z_1^8 - 14z_1^4z_2^4 + z_2^8. \end{cases}$$

We easily verify that, setting aside a numerical factor which occurs, W can also be computed as the Hessian of t.

We obtain a new octahedral form by now constructing the

* We arrive, of course, at the same result if we in general interpret geometrically on the sphere the double ratio of 4 complex values $z = x + iy$, in the way that *Herr Wedekind* has done in his inaugural dissertation (Erlangen, 1874), and in his note on the subject in the Mathematische Annalen (Bd. ix, 1875).

functional determinant of t and W. We thus have, disregarding a factor:

$$(52)\qquad \begin{cases} \chi = z_1^{12} - 33z_1^8z_2^4 - 33z_1^4z_2^8 + z_2^{12}, \text{ or} \\ \chi' = z_1^{12} + 33z_1^8z_2^4 - 33z_1^4z_2^8 - z_2^{12}. \end{cases}$$

We easily prove that this χ is the third special ground-form of the octahedron, *i.e.*, when equated to zero it represents the 12 mid-edge points of the octahedron. In fact, $\chi = 0$ must represent a group of only 12 points connected by means of the octahedral rotations, and since χ is different from t^2 and the group of 6 octahedral points counted twice does not therefore come under consideration, there is, in fact, no other possible explanation.

We have just seen that t and W remain entirely unaltered by the homogeneous tetrahedral substitutions. The same is consequently true of χ. For χ being a covariant, can only alter by a power of the substitution-determinant at the most, if its ground-form is unaltered; but this determinant is in our case equal to 1. Now in § 5 we generated the homogeneous octahedral substitutions by entertaining, in addition to the tetrahedral substitutions mentioned, a single substitution (23) which corresponded to a rotation V of period 4. We determine by direct calculation that t changes its sign for this substitution (and therefore generally for all octahedral substitutions which are not at the same time tetrahedral substitutions). Accordingly W as the Hessian, and since we are again concerned with a substitution of determinant 1, remains generally unaltered, while χ changes its sign alternately just like t, so that the product χt remains unaltered. Thus in any case, t^4, W^3, χ^2, are in general not altered by our homogeneous octahedral substitutions, and there exists, therefore, between them the supposed linear relation. Again, taking into consideration only certain terms in the explicit expressions which result for these forms from (51) and (52), we get for them:

$$(53)\qquad 108t^4 - W^3 + \chi^2 = 0,$$

a relation which holds also for t', W', and χ'.

The form t has been long known in the invariant theory of binary forms, inasmuch as it presented itself as the covariant of the 6th degree of the binary form of the 4th order, when the latter was assumed to be of the canonical form:

$$a(z_1^4+z_2^4)+6bz_1^2z_2^2.$$

Similarly, the synthetic geometers have on many occasions closely investigated the system of points $t=0$, *i.e.*, in their language: the aggregate of 3 mutually harmonic pairs of points. *Clebsch*, too, in his theory of binary algebraic forms, has considered the form t as a special case of general binary forms of the 6th order.* Finally, as regards the relation (53), this, with those analogous to it, are included under a general formula of the theory of invariants, by virtue of which the square of a functional determinant of two covariants is expressed by integral functions of forms of a lower degree.

§ 13. The Set of Forms for the Icosahedron.

To establish the form of the 12th degree, which, equated to zero, represents the 12 summits of the icosahedron, we first calculate the arguments of the several summits, supported by our former developments (§ 6). One of the summits has the argument $z=0$; introducing this into the 60 non-homogeneous icosahedral substitutions (32), we obtain for the 12 summits:

(54) $\quad z=0, \quad \infty, \quad \epsilon^\nu(\epsilon+\epsilon^4), \quad \epsilon^\nu(\epsilon^2+\epsilon^3), \quad (\nu=0, 1, 2, 3, 4).$

We can therefore take the required form f equal to the following product:

$$z_1z_2\cdot\prod_\nu\left(z_1-\epsilon^\nu(\epsilon+\epsilon^4)\,.\,z_2\right)\cdot\prod_\nu\left(z_1-\epsilon^\nu(\epsilon^2+\epsilon^3)z_2\right),$$

or:

$$z_1z_2\left(z_1^5-(\epsilon+\epsilon^4)^5\,.\,z_2^5\right)\left(z_1^5-(\epsilon^2+\epsilon^3)^5\,.\,z_2^5\right);$$

or finally:

(55) $$f=z_1z_2(z_1^{10}+11z_1^5z_2^5-z_2^{10}).$$

We will now again calculate from the f so obtained, discarding the proper numerical factor, the Hessian form, and from this and f calculate the functional determinant. We thus obtain the two forms:

* *Cf.* p. 447, &c. Consult, too, Brioschi, "Sulla equazione del ottaedro," Transunti della Accademia dei N. Lincei 3, iii. (1879), or Cayley, "Note on the Oktahedron Function," Quarterly Journal of Mathematics, t. xvi, 1879.

(56)
$$H=+\frac{1}{121}\begin{vmatrix}\frac{\delta^2 f}{\delta z_1{}^2} & \frac{\delta^2 f}{\delta z_1 \delta z_2}\\ \frac{\delta^2 f}{\delta z_2 \delta z_1} & \frac{\delta^2 f}{\delta z_2{}^2}\end{vmatrix}$$
$$=-(z_1{}^{20}+z_2{}^{20})+228(z_1{}^{15}z_2{}^5-z_1{}^5z_2{}^{15})-494z_1{}^{10}z_2{}^{10},$$

(57)
$$T=-\frac{1}{20}\begin{vmatrix}\frac{\delta f}{\delta z_1} & \frac{\delta f}{\delta z_2}\\ \frac{\delta H}{\delta z_1} & \frac{\delta H}{\delta z_2}\end{vmatrix}$$
$$=(z_1{}^{30}+z_2{}^{30})+522(z_1{}^{25}z_2{}^5-z_1{}^5z_2{}^{25})-10005(z_1{}^{20}z_2{}^{10}+z_1{}^{10}z_2{}^{20}),$$

and I assert with regard to them *that* $H=0$ *represents the* 20 *summits of the pentagon-dodecahedron,* $T=0$ *the* 30 *mid-edge points* (*the ends of the* 15 *cross-lines*).

In order to prove this somewhat more completely than was done in the analogous cases of the tetrahedron and octahedron, let us remark, first, that H and T as covariants of f certainly represent 20 and 30 points respectively on the sphere, such that their totality remains unaltered for the 60 icosahedral substitutions. But now the points on the z-sphere arrange themselves in general by virtue of these rotations into sets of 60, and the number of points thus grouped together is lowered then, and only then, and this to 12, 20, 30 respectively, when we have to do with the summits of icosahedron, the pentagon-dodecahedron, and the mid-edge points. An aggregate of points which remains unaltered for the 60 icosahedral substitutions must be a combination of such individual groups of points. The number of points which it contains necessarily admits of being put into the form:

$$\alpha\,.\,60+\beta\,.\,12+\gamma\,.\,20+\delta\,.\,30,$$

where α, β, γ, δ, are integers, and β, γ, δ, give the multiplicities with which the summits of the icosahedron, the pentagon-dodecahedron, and the mid-edge points contribute to the aggregate of points.

Now if, as in the case of $H=0$, this number is equal to 20, or if, as in the case of $T=0$, it is equal to 30, there is in either case only *one* possible determination of α, β, γ, δ, viz., in the first case $\alpha=\beta=\delta=0$, $\gamma=1$, and in the second case $\alpha=\beta=\gamma=0$, $\delta=1$. But this is what we asserted regarding the meaning of $H=0$, $T=0$.

We now investigate the behaviour of f, H, T, towards the homogeneous icosahedral substitutions with reference to the factors that may occur.

Considering only the *generating* substitutions (24), (26), we determine after a short calculation that f remains unaltered for all of them. The same, therefore, holds good for H and T. For we have defined H and T as covariants of f, and the determinant of each substitution (27) is equal to unity. The behaviour of f, H, T, in this connection is thus as simple as possible. There exists, therefore, certainly, as was supposed above, a linear identity between f^5, H^3, T^2. Again, recurring only to the initial terms of the explicit formulæ (55), (56), (57), we find for this identity:

$$T^2 = -H^3 + 1728 f^5. \tag{58}$$

We have thus found results which are quite analogous to those developed in the case of the tetrahedron and octahedron. If we are to demonstrate here also relations to the general theory of the invariants of binary forms, we cannot at any rate appeal to older works. For the knowledge of the forms f, H, T, was, in fact, first obtained by the consideration of the regular solids and the circumscribed $(x+iy)$-sphere. I first investigated on this basis the principal invariantive properties of the form f in Bd. 9 of the Annalen (l. c.). But there is a series of later publications on the theory of invariants.

These are in connection with the *definition, in the theory of invariants, of the form* f, and of the other forms respectively, which we are considering. In this respect I had myself already announced in Bd. 9 of the Annalen the theorem that f, like the earlier forms Φ and t, is characterised by the identical evanescence of the 4th transvectant $(f, f)^4$. This theorem *Herr Wedekind* had expanded in his "Habilitationsschrift," by showing that, apart from trivial exceptions, in general there is no other binary form whose 4th transvectant with respect to itself vanishes identically except Φ, t, and f.* *Herr Fuchs* has brought forward another property, ana-

* Studien im binären Werthgebiet," Carlsruhe, 1876. See, too, Brioschi, "Sopra una classe di forme binarie," Annali di Matem., 2, viii, 1877. Latterly Brioschi has also considered such forms of the 8th order as are identical save as to a factor with their 4th transvectant. See Comptes Rendus, t. 96 (1883).

logous to this, in his search for these forms,* viz., *that all covariants of these forms which are of a lower degree than the forms themselves, or are powers of forms of a lower degree, must vanish identically. Herr Gordan* then showed † that the property which underlies this is, in fact, just sufficient to characterise the form Φ, t, f. I mention, finally, the latest work of *M. Halphen.*‡ He starts, generally speaking, from the necessity for identities of 3 terms:

$$\lambda_1^{(0)} F_1^{\nu_1} + \lambda_2^{(0)} F_2^{\nu_2} + \lambda_3^{(0)} F_3^{\nu_3} = 0,$$

and shows that these cannot occur otherwise than in the cases which we have investigated. We can thus even regard our forms as defined by these identities. These developments of *M. Halphen* are, moreover, closely related to the others which we shall introduce in the fifth chapter of the present part, when our business is to establish generally all finite groups of binary homogeneous substitutions.

§ 14. The Fundamental Rational Functions.

Having now spent sufficient time over the invariant forms which belong to the homogeneous substitution groups, it is easy to take the final step and construct such rational functions of $z = \frac{z_1}{z_2}$ as remain in general unaltered by the non-homogeneous substitutions of § 7. In fact, we shall only have to establish proper quotients of our invariant forms of null dimensions in z_1 and z_2. We asserted in § 1 that in all cases *one* such quotient Z could be constructed, which, equated to a constant, uniquely represents in each case the different groups of points on the sphere such as we are considering. This is clearly nothing less than saying that there exists a rational function of the kind required which is of degree N, understanding by N the number of the non-homogeneous sub-

* See the Göttinger Nachtrichten of December 1875, as also the memoirs in Borchardt's Journal, Bd. 81, 85 (1876–78). The "Primformen," which Herr Fuchs there considers, are just what we have called in the text "groundforms."

† Math. Ann., Bd. xii (1877): "Bin. Formen mit versch. Covarianten."

‡ "Mem. présentés par divers savants à l'Académie," &c., t. 28 (1883): "Mémoire sur la réduction des équations diff. lin. aux formes intégrables" (Prize-essay of the Paris Academy, 1880).

stitutions in question. Before we actually establish these *fundamental rational functions*, and thus provide the shortest proof of their existence, it will be useful to make inquiries as to their position among the other rational functions which remain unaltered.

I say first *that every such rational function of* z *is a rational function of* Z. In fact, if $R(z)$ be such a function, $R(z)$ will assume the same value for all points on the sphere which proceed from it by means of the N rotations of the group in question, but the N points so connected are, by hypothesis, characterised by one value of Z. The functions Z and R, which, through the intervention of z, are always algebraical functions of one another, are therefore so related that to every value of Z only *one* value of R corresponds, *i.e.*, R is a rational function of Z, *q.e.d.* That conversely every rational function of Z is a function $R(z)$, scarcely needs mentioning.

I say further, *that, by the property attributed to it,* Z *is fully determined save as to linear transformations*, viz., let Z' be a second rational function of z, which, like Z, has the property of representing, when equated to a constant, only one group of connected points. We conclude, just as before, that Z' depends rationally on Z, but that also Z depends rationally on Z'. Therefore Z' is a *linear* function of Z: $Z'=\frac{\alpha Z+\beta}{\gamma Z+\delta}$. It is again manifest that we should be able conversely to use every Z' introduced in this way as our fundamental rational function just as well as the original Z.

On the last remark is based the following: *that we can subject our fundamental rational function* Z *to three more independent conditions*, to make it fully determinate. First with regard to the *cyclic groups*, we simply put:

$$Z=\left(\frac{z_1}{z_2}\right)^n, \tag{59}$$

where Z therefore vanishes for one pole of the cyclic group, and becomes infinite for the other, and takes along the equator the absolute measurement unity. In the other cases, we have always, as we know, to distinguish three special groups of points, which, with the multiplicities ν_1, ν_2, ν_3 respectively, are contained within the general groups of points appertaining thereto. Following a method frequently employed, we now so regulate our Z that it assumes for these three groups of points the values 1, 0, ∞, respectively. Then Z will take the form $\frac{c\,.\,F_2^{\nu_2}}{F_3^{\nu_3}}$, and $Z-1$ the analogous form $\frac{c'\,.\,F_1^{\nu_1}}{F_3^{\nu_3}}$, where by F_1, F_2, F_3 are to be understood what we have previously called the ground-forms. At the same time, c and c' must be of such a nature that the equation:

$$c\,.\,\frac{F_2^{\nu_2}}{F_3^{\nu_3}}-1=c'\,.\,\frac{F_1^{\nu_1}}{F_3^{\nu_3}},$$

coincides with the oft-mentioned identity existing between F_1, F_2, F_3, which fully determines c and c'.

Turning now to the task of giving explicitly in every case the function Z thus defined, I make use of a notation which uniformly connects the two expressions of Z and $Z-1$, viz., I put $Z:Z-1:1$ proportioned to:

$$cF_2^{\nu_2}:c'F_1^{\nu_1}:F_3^{\nu_3}.$$

We obtain in this form the following table, to which we shall often recur:

(1.) *Dihedron:*

$$(60)\qquad Z:Z-1:1=\left(\frac{z_1^n-z_2^n}{2}\right)^2:\left(\frac{z_1^n+z_2^n}{2}\right)^2:-(z_1z_2)^n;$$

(2.) *Tetrahedron:*

$$(61a)\qquad Z:Z-1:1=\Psi^3\ :\ -12\sqrt{-3}\,.\,t^2\ :\Phi^3,$$

or

$$(61b)\qquad Z:Z-1:1=\Psi'^3:\ -12\sqrt{+3}\,.\,t'^2:\Phi'^3,$$

according as we assume the first or second position of the system of co-ordinates.

(3.) *Octahedron*, with the same distinction:

$$(62a)\qquad Z:Z-1:1=W^3\ :\chi^2\ :108t^4,$$

or

$$(62b)\qquad Z:Z-1:1=W'^3:\chi'^2:108t'^4;$$

(4.) *Icosahedron:*

$$(63)\qquad Z:Z-1:1=H^3\ :\ -T^2\ :1728f^5.$$

For the symbols here applied, consult throughout the principal formulæ of paragraphs 11, 12, and 13.

§ 15. Remarks on the Extended Groups.

Finally, we return to our extended groups (§ 7) once more. We want to know how our rational fundamental functions now obtained behave towards them. From the analytical side the extended groups l. c. arose from a combination of the operation $z' = \bar{z}$ with the non-homogeneous groups of substitutions, where, so far as the tetrahedron was concerned, we only supposed the second position of the co-ordinate system to be employed. But now, maintaining the same supposition, *all* our ground-forms have real coefficients, and Z will be derived from these ground-forms, in virtue of the preceding formulæ, in every case by the help of real coefficients. The matter therefore simply comes to this: *that for all those operations of the extended groups which are not already contained in the corresponding non-homogeneous groups of substitutions, Z in each case passes over to its conjugate imaginary value.*

Combining this result with the propositions which we deduced in § 11 of the preceding chapter, we obtain one final remarkable result. It is this: *Z assumes real values for all those points for the z-sphere which lie in the planes of symmetry of the configuration in question, and only for such points.* The points of the said planes of symmetry are therefore in each case characterised by the reality of the corresponding Z.

Looking back, we have in the second chapter thus ended arrived at this point: we have connected the geometrical results of the group-theory occurring in the first chapter with a definite region of recent mathematic, namely, with the *algebra of linear substitutions* and the corresponding *theory of invariants.* Just in the same way, the following two chapters are destined to effect the connection with the two other modern theories. These are *Riemann's theory of functions* and *Galois's theory of algebraical equations.*

CHAPTER III

STATEMENT AND DISCUSSION OF THE FUNDAMENTAL PROBLEM, ACCORDING TO THE THEORY OF FUNCTIONS

§ 1. Definition of the Fundamental Problem.

The investigations of the preceding chapter have led us, in the formulæ (59)—(63) of the last paragraph but one, to the knowledge of certain rational functions Z of z, which remain unaltered for the groups of non-homogeneous substitutions in each case considered, and by means of which all other rational functions of z, which remain unaltered, are expressed rationally. To this result we add the statement of a relation which we denote as the *equation* appertaining to the group in each case. *We suppose, namely, that the numerical value of Z is arbitrarily given, and seek to calculate from it the corresponding z as the unknown;* or, to express it differently: *we no longer consider Z as a function of z, but z as a function of Z.* The equation which thus corresponds to the cyclic group is, according to formula (59), l. c., none other than the *binomial* equation:

$$\left(\frac{z_1}{z_2}\right)^n = Z. \tag{1}$$

The other equations correspond in just the same way to the formulæ (60—63). I will collect them here briefly in the form:

$$c \cdot \frac{F_2^{\nu_2}}{F_3^{\nu_3}}, \tag{2}$$

which we used incidentally in the preceding chapter. Here F_2, F_3, together with F_1, denote those three principal forms of which all other invariant forms are compounded as integral functions, and ν_2, ν_3 are in each case taken from the table which was provided in § 9 of the preceding chapter, and which I reproduce here to facilitate reference:

(3)

	ν_1	ν_2	ν_3	N
Dihedron . . .	2	2	n	$2n$
Tetrahedron . .	2	3	3	12
Octahedron . .	2	3	4	24
Icosahedron . .	2	3	5	60

I have here added a last column, headed by N, which marks the *degree* of the equation in each case under consideration.*

But with the equations (1), (2), only a part of our earlier considerations is inverted; we obtain a second mode of presenting the problem by recurring to the several invariant forms themselves. These forms remain unaltered by the homogeneous substitutions of determinant 1 in general, save as to a factor. It is not difficult, however, to select from them those for which this factor is equal to 1, and which we can call the *absolute* invariants. The sequel shows that these absolute invariants can be composed in every case as integral functions of three of them; I have noted these three forms in the following table, together with the identities subsisting between them in each case:

I. *Cyclic groups.*

(4) $\begin{cases} \text{Forms: } z_1 z_2,\ z_1^{2n},\ z_2^{2n}; \\ \text{Identity: } (z_1 z_2)^{2n} = z_1^{2n} \cdot z_1^{2n}. \end{cases}$

II. *Dihedral groups.*

In the case of the dihedron we had:

$$F_1 = \frac{z_1^{\,n} + z_2^{\,n}}{2}, \quad F_2 = \frac{z_1^{\,n} - z_2^{\,n}}{2}, \quad F_3 = z_1 z_2,$$

and the relation:

$$F_1^{\,2} = F_2^{\,2} + F_3^{\,n}.$$

* I shall also occasionally denote the degree of (1) by N in the following pages.

If we now seek the absolute invariants, we obtain for *n even:*

(5a) $\begin{cases}\text{Forms:} & F_3^2,\ F_1^2,\ F_1F_2F_3;\\ \text{Identity:} & (F_1F_2F_3)^2 = F_1^2 \cdot F_3^2 \cdot (F_1^2 - F_3^n);\end{cases}$

and for *n odd:*

(5b) $\begin{cases}\text{Forms:} & F_3^2,\ F_1^2F_3,\ F_1F_2;\\ \text{Identity:} & (F_1F_2)^2 \cdot F_3^2 = (F_1^2F_3) \cdot (F_1^2F_3 - F_3^{n+1}).\end{cases}$

III. *Tetrahedral group:* *

(6) $\begin{cases}\text{Forms:} & F_1 = t,\ F_2F_3 = W,\ F_2^3 = \Phi^3;\\ \text{Identity:} & W^3 = \Phi^3(\Phi^3 - 12\sqrt{-3} \cdot t^2).\end{cases}$

IV. *Octahedral group:*

(7) $\begin{cases}\text{Forms:} & F_2 = W,\ F_3^2 = t^2,\ F_1F_3 = \chi t;\\ \text{Identity:} & (\chi t)^2 = t^2(W^3 - 108t^4).\end{cases}$

V. *Icosahedral group:*

(8) $\begin{cases}\text{Forms:} & F_1 = T,\ F_2 = H,\ F_3 = f;\\ \text{Identity:} & T^2 + H^3 - 1728f^5 = 0.\end{cases}$

We now suppose, in a particular case, that the numerical value of the three forms included in the table, in correspondence with the identity subsisting between them, is given, and we seek *to calculate from this the values of the two variables* z_1, z_2. Thus we have what we will call the form-problem. The number of the systems of solution of a form-problem is always $2N$, where by N is to be understood the degree of the corresponding equation. All these systems of solution proceed in this case, in just the same way, from any one of them in virtue of the $2N$ homogeneous substitutions, as the N solutions of each equation manifestly do with respect to the N non-homogeneous substitutions.

§ 2. Reduction of the Form-Problem.

As regards the solution of the form-problem, we can always accomplish it by means of the corresponding equation and an accessory square root. Take, for instance, the cyclic groups. We then calculate first from the forms (4) the right side of (1):

$$Z = \frac{(z_1z_2)^n}{z_2^{2n}} = \frac{z_1^{2n}}{(z_1z_2)^n},$$

* In the case of the tetrahedron and octahedron, I now use, contrary to what I have hitherto done, non-accented letters.

then solve (1), whence we find $\frac{z_1}{z_2}=z$, and finally obtain z_1, z_2 themselves by introducing this value of $\frac{z_1}{z_2}$ into the given form of the second degree $z_1 z_2$ (which we shall now call X), whence:

$$(9) \qquad z_2=\sqrt{\frac{X}{z}},\ z_1=z \cdot z_2.$$

In the case of the other groups, the matter takes a form perfectly analogous. For not only does the particular Z (2) in these cases also admit of being rationally composed of the forms (5)—(8), but we can also always construct rationally from these forms an expression which is of the second degree in z_1, z_2. I choose as such, in all the cases:

$$(10) \qquad X=\frac{F_2 \cdot F_3}{F_1}.$$

If we have then determined, by means of (2), the quotients $\frac{z_1}{z_2}=z$, we find, by comparison with (10):

$$(11) \qquad z_2=\sqrt{\frac{X(z_1, z_2)}{X(z, 1)}},\ z_1=z \cdot z_2,$$

where $X(z_1, z_2)$ denotes the magnitude (10) previously given, and $X(z, 1)$ a definite rational function of z:

$$\frac{F_2(z, 1) \cdot F_3(z, 1)}{F_1(z, 1)}.$$

We have thus at the same time the means of simplifying the previous statement of our form-problem, of *reducing* it, as we will say.* By means of (9) and (11), z_1, z_2 depend only on X and Z, which, in their turn, are rational functions of the forms (4)—(8). We now introduce these values of z_1, z_2 into the forms (4)—(8). Thus these forms will be rational in X, since they are all of even degree. *But at the same time they will be also rational in Z.* For they now represent rational function of z, such as do not alter for the N corresponding non-homo-

* That such a reduction was possible was pointed out to me incidentally by *Herr Nöther*, who derived it in a totally different manner from his general researches on the conformable representation of surfaces.

geneous substitutions. We shall therefore in the sequel, when speaking of the form-problems, not suppose, say, the forms (4)—(8) to be given [where we had always to pay regard to the identities subsisting between them], *but rather the expressions Z and X primarily*, and then consider z_1, z_2 as functions of these two magnitudes.

I reproduce here explicitly the rational functions of Z and X, to which the forms (4)—(8) are equal. We verify these easily by reflecting, on the one hand, how Z and X are composed of the forms (4)—(8), and, on the other hand, taking account of the identities subsisting between these forms. I find:

I. *For the cyclic groups:*

$$z_1 z_2 = X,\quad z_1^{2n} = Z \, . \, X\,,\quad z_2^{2n} = \frac{X}{Z}\,. \tag{12}$$

II. *For the dihedron:* for *n even:*

$$F_3^2 = \frac{X^2 \, . \, Z-1}{Z},\quad F_1^2 = -\frac{X^n \, . \, (Z-1)^{\frac{n+2}{2}}}{Z^{\frac{n}{2}}}, \tag{13a}$$

$$F_1 F_2 F_3 = -\frac{X^{n+1} \, . \, (Z-1)^{\frac{n+2}{2}}}{Z^{\frac{n}{2}}},$$

and, for *n odd:*

$$F_3^2 = \frac{X^2 \, . \, Z-1}{Z},\quad F_1^2 F_3 = -\frac{X^{n+1} \, . \, (Z-1)^{\frac{n+3}{2}}}{Z^{\frac{n+1}{2}}}, \tag{13b}$$

$$F_1 F_2 = -\frac{X^n \, . \, (Z-1)^{\frac{n+1}{2}}}{Z^{\frac{n-1}{2}}}.$$

III. *For the tetrahedron:*

$$F_1 = -\frac{X^3 \, . \, (Z-1)^2}{432Z},\quad F_2 F_3 = -\frac{X^4 \, . \, (Z-1)}{432Z}, \tag{14}$$

$$F_2^3 = -\frac{X^6 \, . \, (Z-1)^3}{5184\sqrt{-3} \, . \, Z}.$$

IV. *For the octahedron :*

$$(15)\qquad F_2 = 108\,.\,\frac{X^4\,.\,(Z-1)^2}{Z},\ F_3{}^2 = 108\,.\,\frac{X^6\,.\,(Z-1)^3}{Z^2},$$

$$F_1F_3 = 108^2\,.\,\frac{X^9\,.\,(Z-1)^5}{Z^3}.$$

V. *For the icosahedron :*

$$(16)\qquad F_1 = 12^9\,.\,\frac{X^{15}\,(Z-1)^8}{Z^5},\ F_2 = -12^6\,.\,\frac{X^{10}\,(Z-1)^5}{X^3},$$

$$F_3 = -12^3\,.\,\frac{X^6\,(Z-1)^3}{Z^2}.$$

§ 3. Plan of the Following Investigations.

We have now to discuss the fundamental problems, which we have thus far reached, under a double aspect, viz., in the sense of the theory of functions, and algebraically. Postponing the latter kind of investigations to the following chapter, let us turn at once to the function-theory considerations.

We have z, the unknown in the particular *equation*, as a function of Z alone, while the z_1, z_2 of the corresponding *form-problem* depends also on X. But the mode of dependence by formulæ (9) and (11) is so extremely simple that we need delay no longer over it. We will, therefore, only discuss z_1 and z_2 so far as they are functions of Z.

Such an investigation divides itself naturally into two parts. We have first to obtain a *general survey* of the different branches of our functions, and then to suggest the means of *computing the particular branch of the function* by a convergent process (for example, by a series of powers). We attain the former very simply, in our case, by the method of conformable representation (§§ 4, 5). We learn hereby, at the same time, the *form* of the series which come under consideration for the different branches of our functions (§ 5). The coefficients of the expansions will then be given by proving *that z satisfies, in relation to Z, a simple differential equation of the third order, and consequently the roots z_1, z_2 of the parallel form-problem appear as solutions of a homogeneous linear differential equation of the second order, with rational coefficients* (§§ 6–9). Finally, we prove in § 10 that, by reason of the last-mentioned differential equation, z_1, z_2 are particular cases of Riemann's P-function, whereupon our investigations are seen to join a well-defined and much-explored region of modern analysis.

As to the results which we obtain in this way, they are, in their main features, all contained already in the above-mentioned work of *Herr Schwarz*;* except that in Herr Schwarz's article the order of the matter is just the reverse of that followed by us here. Starting from the differential equation of the hypergeometric series, Herr Schwarz first constructs the differential equation of the third order, on which the quotient z of two particular solutions z_1, z_2 depends. He then investigates the conformable representation, which z effects, of the two half-planes of the independent variable Z, and proceeds finally, by means of the condition that z is to be an *algebraical* function of Z, to the z-functions considered by us and the fundamental equations which define them.† We, on the contrary, begin with these equations, construct from them the conformable representation, and then deduce the existence of the differential equations of the third order, which z satisfies, and, finally, pass from this to the differential equation of the second order of the P-function, or, what is essentially the same, of the hypergeometric series. In this connection it may be here explained that, in taking this last step, we borrow an idea which *Herr Fuchs* has introduced in his memoirs mentioned above,‡ inasmuch as we represent $X(z_1, z_2)$ (a *form*, therefore, dependent on z_1, z_2) directly by means of Z.

I should, of course, have been able to collect the developments here described much more briefly had I desired to presuppose special knowledge with regard to Riemann's P-function, or even merely to make use of the general foundations of the modern theory of linear differential equations with rational coefficients, as developed by Herr Fuchs § in the 66th volume of Borchardt's

* "Ueber diejenigen Fälle, in welchen die Gaussische hypergeometrische Reihe eine algebraische Function ihres vierten Elementes darstellt." Borchardt's Journal, Bd. 75, pp. 292–335 (1872).

† I summarise in the text only such of the results obtained by Herr Schwarz as are in immediate relation with our own exposition.

‡ See the reference on p. 63.

§ "Zur theorie der linearen Differentialgleichung mit veränderlichen Coefficienten" (1865).

Journal. This sacrifice being made, my exposition has the advantage of leading, by a relatively short route, to a portion of the researches just mentioned. I should like to refer here in this relation to § 3 of the fifth chapter following, where, in connection with the development now given, the most general linear differential equations of the second order with rational coefficients, and which have entirely algebraical integrals, are directly determined.

§ 4. On the Conformable Representation by Means of the Function $z\ (Z)$.

Turning now to the conformable representation which is furnished by $z\ (Z)$, we denote as before the complex values of $z=x+iy$ on the sphere, while we interpret $Z=X+iY$ on a plane.* We construct in the plane Z the axis of real numbers, and divide this into a *positive* and *negative half-plane*. We mark in addition, when we have to do with the binomial equations (1), the two points $Z=0,\ \infty$, in the other cases the three points $Z=1,\ 0,\ \infty$.

A glance at the equations (1), (2), and again at the more complete formulæ (59)—(63) of the preceding chapter, teaches us that, in the case of the binomial equations, the n function branches coming under consideration for $Z=0$ and $Z=\infty$ all congregate in cycle, while, in the other cases, for $Z=1$, ν_1 of the N existing branches are connected cyclically; for $Z=0$, ν_2; and for $Z=\infty$, ν_3. *Now I say that the function $z\ (Z)$ furnishes no other branchings than those given here.* In general, viz., when Z is given as a rational function of $z=\frac{z_1}{z_2}$ in the form:

$$Z=\frac{\phi\ (z_1,\ z_2)}{\psi\ (z_1,\ z_2)},$$

[where ϕ, ψ, are to be integral homogeneous functions of the accompanying argument, of degree N], we find those values of z, and therefore of Z, for which branchings take place, by

* Whoever is not thoroughly familiar with the theory of the conformable representation will consult with advantage Herr Holzmüller's recently published work, "Einführung in die Theorie der isogonalen Verwandtschaft und der conformen Abbildungen," &c. (Leipzig, 1882).

equating to zero the functional determinant of the $(2N-2)^{\text{th}}$ degree:

$$\frac{\delta\phi}{\delta z_1}\cdot\frac{\delta\psi}{\delta z_2}-\frac{\delta\psi}{\delta z_1}\cdot\frac{\delta\phi}{\delta z_2}.$$

If this vanishes μ-times at a position $z=z_0$, $\mu+1$ branches of the function z for $Z=Z_0$ are connected cyclically in correspondence therewith.* If we compute this functional determinant in any one of our cases (1), (2), we always return to the branching points, which we already know. For in the case of the binomial equations we obtain simply:

$$z_1^{n-1}\,.\,z_2^{n-1}=0;$$

and in the case of the other equations, recalling that ν_1 is always $=2$, and F_1 is the functional determinant of F_2 and F_3:

$$F_1^{\nu_1-1}\,.\,F_2^{\nu_2-1}\,.\,F_3^{\nu_3-1}=0,$$

where the different roots of $F_1=0$ all give $Z=1$, those of $F_2=0$, $Z=0$, and finally those of $F_3=0$, $Z=\infty$.†

The data so attained are already sufficient to characterise fully the nature of the conformable representation which we sought. If we describe as an n-gon every figure situated on the sphere, and furnished with the necessary number of summits, and otherwise bounded by continuously curved lines, and observe that Z is rational in z, and that therefore to every Z belong N values of z, while to every z belongs only one value of Z, we have at once:

In virtue of the binomial equation (1), *the two half-planes Z will be alternately represented on $2N$ lunes of the z-sphere which meet at the poles of the z-sphere* (i.e., *the points* $z_1z_2=0$) *with*

* The rule here formulated differs from that given in the text-books in the use of the homogeneous variables z_1, z_2. This has the advantage of embracing in one form of expression the finite and infinite values of z, as the geometrical interpretation of z on the sphere and the modern conception generally of the infinite requires.

† This explicit calculation of the functional determinant was not really needed for the establishment of our result; it would have been sufficient to have remarked that the total number of the branching points for $Z=0$, ∞, and for $Z=1$, 0, ∞, respectively (with their proper multiplicities taken into account) is identical with the degree $(2N-2)$ of the functional determinant. [We must here attribute $(\nu-1)$ roots of the functional determinant in each case to ν branches associated in cycle.]

angles $=\frac{\pi}{N}$, *and envelope the z-sphere completely, but nowhere multiply.*

Just in the same way in the cases (2), *the half-planes Z will be represented alternately on 2N triangles of the z-sphere, which, with angles equal to* $\frac{\pi}{\nu_1}, \frac{\pi}{\nu_2}, \frac{\pi}{\nu_3}$, *extend to one point of* $F_1=0$, *one point of* $F_2=0$, *and one point of* $F_3=0$.

We now observe that all roots of (1) or (2) are successively derived from any one of themselves, in each case, by N linear substitutions to which correspond *rotations* of the z-sphere round the centre. We thus conclude immediately that:

The N lunes or triangles which in an individual case correspond to the positive half-plane Z, as also the N lunes or triangles which correspond to the negative half-plane Z, are respectively congruent with one another.

Finally, we recall the theorem which we deduced in the concluding paragraph of the preceding chapter from the existence of the extended group. We there showed that Z only assumes real values along those great circles of the z-sphere which are traced out by the planes of symmetry of the several configurations. Now the real values of Z separate in the Z-plane the two half-planes. Hence we have finally:

The boundary lines of the lunes and triangles are none other than the circles of symmetry before mentioned, and our lunes and triangles are therefore identical with those figures which we have described in § 11 *of the first chapter as fundamental domains of the extended group.*

I beg the reader to make himself quite familiar with the formal relations here described; this is not the place to discuss them more minutely.* The representation which corresponds to the binomial equations has of course been much investigated elsewhere, only that the z-sphere has been replaced throughout by the plane to which we must suppose our sphere related by means of stereographic projection.†

* As regards the icosahedral equation in particular, a glance at the figure gives the elegant theorem: *that this equation, for a real value of Z, possesses always four, but only four, real roots.*

† In his "Vorlesungen über mathematische physik" (Leipzig, 1876), *Herr Kirchoff* describes those plane figures which correspond to our lunes as *Sicheln.*

For the rest, I will in the developments of the following paragraphs leave on one side the binomial equations and the cyclic groups generally, in consideration of the gap which separates them from the other cases, and only note the simple results which relate to them in footnotes.

§ 5. March of the z_1, z_2 Function in General—Development in Series.

The characteristic feature of the geometrical expression of the functions $z\ (Z)$, as we have given it in the preceding paragraphs, consists in the fact that we have constructed, not a many-leaved surface on the Z-plane, but a region-partition on the z-sphere.* Having now to consider the march of the functions $z_1\ (Z)$, $z_2\ (Z)$, we transfer our attention, accordingly, again to the z-sphere. Leaving aside, as proposed, the cyclic groups, we have to recur to the formulæ (11), which we will write in the following manner:

(17) $$z_2 = \sqrt{X \cdot \frac{F_1(z,\,1)}{F_2\,(z,\,1)\,.\,F_3\,(z,\,1)}},\ z_1 = z\,.\,z_2.$$

Here z_1, z_2 appear as single-valued functions of position on a two-leaved surface, covering the z-sphere, which possesses branch-points at all points $F_1 = 0$, or $F_2 = 0$, or $F_3 = 0$ (the point $z = \infty$ not excluded), and therefore belongs to the deficiency:

(18) $$p = -1 + \frac{N}{2}\left(\frac{1}{\nu_1} + \frac{1}{\nu_2} + \frac{1}{\nu_3}\right).$$

We determine at once for the particular function its null and infinite points, which of course must occur in equal numbers. As concerns z_2, it vanishes, and in fact *simply* † vanishes, for all points of $F_1 = 0$, and also for $z = \infty$, on the whole, therefore, for $\left(\frac{N}{\nu_1} + 1\right)$ points. On the other hand, it becomes simply

* In a similar manner, the march of any one-valued function $Z = F(z)$ can be exhibited. *Cf.* for example, *O. Hermann:* "Geometrische Untersuchungen über den Verlauf der elliptischen Transcendenten im complexen Gebiete," Schlömilch's Zeitschrift, Bd. 28 (1883).

† We say of a function which becomes zero or infinite at a branch-point z_0 on a two-leaved surface, that it becomes *simply* zero or infinite, if it behaves for a first approximation like $C(z - z_0)^{\frac{1}{2}}$ or $C(z - z_0)^{\frac{1}{2}}$ respectively. If $z_0 = \infty$, we have to consider instead of $(z - z_0)$ the expression $\frac{1}{z}$.

infinite for all points of $F_2=0$, and those points of $F_3=0$ which do not coincide with $z=\infty$; the number of the infinite points is therefore $\left(\frac{N}{\nu_2}+\frac{N}{\nu_3}-1\right)$, which, in fact, is identical with $\left(\frac{N}{\nu_1}+1\right)$ for the numerical value of N and ν we are considering. Just in the same way for z_1, only that the two points $z=0$ and $z=\infty$ (which both belong to the roots of $F_3=0$) have exchanged places.

We can now with little trouble display the nature of the development in series of which our three functions z, z_1, z_2 admit in the neighbourhood of the singular positions $Z=1, 0, \infty$. I only complete this here so far as we use it in the following paragraphs. Let us agree for a moment (as is indeed otherwise customary) that $Z-Z_0$ shall denote the value $\frac{1}{Z}$ for $Z_0=\infty$ and correspondingly $z-z_0$ the value $\frac{1}{z}$ for $z_0=\infty$. Further, let z_0 be one of the values of z which belong to $Z=Z_0$. Then we have directly, from the conformal representation of the preceding paragraph, the following general theorem:

In the neighbourhood of $Z_0=1, 0, \infty$, $z-z_0$ *admits of a development in an ascending series of powers:*

$$(19) \qquad z-z_0=a\,(Z-Z_0)^{\frac{1}{\nu}}+b\,(Z-Z_0)^{\frac{2}{\nu}}+\ldots$$

where ν *is to denote the numbers* ν_1, ν_2, ν_3, *in order, and the coefficient* a *is different from zero.*

We consider now in particular the case $Z_0=\infty$, $z_0=0$, and the corresponding developments of z_1, z_2. The formula (2) to which we must here return:

$$c\cdot\frac{F_2^{\nu_2}}{F_3^{\nu_3}}=Z,$$

contains on the left-hand side the factor $\frac{c}{z^{\nu_3}}$ multiplied by a rational function of z^{ν_3}, which for $z=0$ assumes the value $+1$, and for the icosahedron the value -1. Hence we have first for z the development:

$$(20) \qquad z=\left(\frac{\pm c}{Z}\right)^{\frac{1}{\nu_3}}\cdot\mathfrak{P}\left(\frac{1}{Z}\right),$$

where the minus sign only occurs in the case of the icosahedron and $\mathfrak{B}\left(\frac{1}{Z}\right)$ denotes a series proceeding according to integral powers of $\frac{1}{Z}$ of which the first coefficient is equal to $+1$. We consider now the formulæ (17). The quotient occurring in them $\frac{F_1(z, 1)}{F_2(z, 1) \cdot F_3(z, 1)}$ breaks up into the product $\frac{1}{z}$ and a rational function of z^{ν^3}, which again for $z=0$ is equal to $+1$, but in the case of the icosahedron is equal to -1. *Introducing now for z the series* (20), *the two minus signs which occur in the case of the icosahedron evidently destroy one another, ν_3 being an odd number in the case of the icosahedron. We obtain from* (20) *and* (17) *the following series for z_1, z_2:*

$$(21) \qquad \begin{cases} z_1 = \sqrt{X} \cdot \left(\frac{c}{Z}\right)^{\frac{1}{2\nu_3}} \cdot \mathfrak{B}_1\left(\frac{1}{Z}\right), \\ z_2 = \sqrt{X} \cdot \left(\frac{Z}{c}\right)^{\frac{1}{2\nu_3}} \cdot \mathfrak{B}_2\left(\frac{1}{Z}\right), \end{cases}$$

where $\mathfrak{B}_1$, $\mathfrak{B}_2$, are series of powers which proceed according to integral powers of $\frac{1}{Z}$ and begin with the term $+1$.

We shall not return to the formulæ thus obtained till § 10. Let us recollect, meanwhile, that c in the case of the dihedron $=-1$, for the tetrahedron $=+1$, while it has for the octahedron the $\frac{1}{108}$, and for the icosahedron the value $\frac{1}{1728}$.

§ 6. Transition to the Differential Equations of the Third Order.

We now turn to the consideration of that differential equation of the third order with rational coefficients which z, as we asserted above, satisfies in relation to Z. This has its origin in the property *that all the N branches of z are linear functions of one of themselves,* and, in fact, in the following way: understanding by η an arbitrary function of Z, let us eliminate generally between $\frac{\alpha\eta+\beta}{\gamma\eta+\delta}$ and its first, second, and third differential coefficients the three constants $\alpha : \beta : \gamma : \delta$. We thus obtain a differential expression of the third order which remains unaltered for any linear transformations of η. Now, substituting our z for η, this differential expression, in virtue of the property just explained of the N functional branches of z, will take a determinate value independent of the branch which we may choose. *Therefore for $\eta = z$ the said differential expression is a one-valued function of Z, and therefore also (since z is algebraic in Z) a rational function of Z.* Putting it equal to the proper rational function of Z, we have the proposed differential equation of the third order, which $\eta = z$ satisfies as a particular solution.

Our first object is to actually construct this differential expression of the third order. Let $\zeta = \frac{\alpha\eta + \beta}{\gamma\eta + \delta}$, or, as we will write it:

$$\gamma\eta\zeta - \alpha\eta + \delta\zeta - \beta = 0,$$

and then on differentiating successively with respect to Z:

$$\begin{aligned} \gamma(\eta'\zeta + \eta\zeta') \quad - \alpha\eta' \ + \delta\zeta' &= 0, \\ \gamma(\eta''\zeta + 2\eta'\zeta' + \eta\zeta'') \quad - \alpha\eta'' - \delta\zeta'' &= 0, \\ \gamma(\eta'''\zeta + 3\eta''\zeta' + 3\eta'\zeta'' + \eta'\zeta''') - \alpha\eta''' - \delta\zeta''' &= 0. \end{aligned}$$

In the three equations thus obtained, β has vanished of itself, the elimination of the other constants gives, after an easy reduction:

$$0 = \begin{vmatrix} 0 & \zeta' & \eta' \\ 2\eta'\zeta' & \zeta'' & \eta'' \\ 3\eta''\zeta' + 3\eta'\zeta'' & \zeta''' & \eta''' \end{vmatrix},$$

or, on separating the variables:

$$\frac{\zeta'''}{\zeta'} - \frac{3}{2}\left(\frac{\zeta''}{\zeta'}\right)^2 = \frac{\eta'''}{\eta'} - \frac{3}{2}\left(\frac{\eta''}{\eta'}\right)^2,$$

The differential expression required is therefore:

$$(22) \qquad \frac{\eta'''}{\eta'} - \frac{3}{2}\left(\frac{\eta''}{\eta'}\right)^2.$$

We will in future denote this by $[\eta]$ or by $[\eta]_z$.* We will,

* According to a communication for which I am indebted to *Herr Schwarz*, this expression occurs in Lagrange's researches on conformable representation: "Sur la construction des cartes géographiques," Nouv. Mem. de l'Acad. de Berlin, 1779. *Cf.* further Herr Schwarz's often-mentioned treatise in Bd. 75 of Borchardt's Journal, where other literary notes are collected. In the "Sitzungsberichten der sächsischen Gesellschaft" of January 1883, I have tried to demonstrate what deeper meaning is involved in a differential equation of the third order $[\eta] = f(z)$ if we start from the origin of the expression $[\eta]$ as it is treated of in the text.

moreover, here estimate how $[\eta]_Z$ varies if we introduce instead of Z a new variable Z_1. If

$$Z = F(Z_1)\ Z' = \frac{dZ}{dZ_1}, \text{ \&c. } \ldots \ldots$$

there follow in order:

$$\frac{d\eta}{dZ_1} = \frac{d\eta}{dZ} \cdot Z',$$

$$\frac{d^2\eta}{dZ_1^2} = \frac{d^2\eta}{dZ^2} \cdot Z'^2 + \frac{d\eta}{dZ} \cdot Z'',$$

$$\frac{d^3\eta}{dZ_1^3} = \frac{d^3\eta}{dZ^3} \cdot Z'^3 + 3\frac{d^2\eta}{dZ^2} \cdot Z'Z'' + \frac{d\eta}{dZ} \cdot Z'''.$$

Therefore

$$(23) \qquad [\eta]_{Z_1} = [\eta]_Z \cdot Z'^2 + [Z]_{Z_1},$$

which is the required formula. If, in particular, Z depends linearly on Z_1,

$$Z = \frac{AZ_1 + B}{CZ_1 + D},$$

then $[Z]_{Z_1}$ disappears, and we have simply

$$(24) \qquad [\eta]_{Z_1} = [\eta]_Z \cdot \frac{(AD - BC)^2}{(CZ_1 + D)^4}.$$

§ 7. Connection with Linear Differential Equations of the Second Order.

Before going further, we will unfold the connection between the said differential equation of the third order and the homogeneous linear differential equations of the second order, which we shall have immediate occasion to utilise. Suppose that, in general, a linear differential equation with rational coefficients is given:

$$(25) \qquad y'' + p \cdot y' + q \cdot y = 0.$$

Understanding by y_1, y_2, any two partial solutions of it, let us put

$$\eta = \frac{y_1}{y_2}.$$

If we then allow Z to describe any closed path in its plane, η will only be able to pass over into a linear function of itself

$\frac{\alpha\eta+\beta}{\gamma\eta+\delta}$. For after any such cycle, y_1, y_2 have only transformed themselves into certain linear combinations of y_1, y_2. *Hence we conclude that our η satisfies a differential equation of the third order of the kind just considered:*

(26) $$[\eta]_Z = r(Z),$$

understanding by $r(Z)$ a rational function of Z.

Our next object must be to calculate this $r(Z)$ in terms of the coefficients p, q of (25). By supposition:

$$y_1'' + p \cdot y_1' + q \cdot y_1 = 0,$$
$$y_2'' + p \cdot y_2' + q \cdot y_2 = 0.$$

therefore combining the two equations:

(27) $$(y_1''y_2 - y_2''y_1) + p(y_1'y_2 - y_2'y_1) = 0.$$

We have further:

(28) $$\frac{y_1'y_2 - y_2'y_1}{y_2^2} = \eta',$$

whence by logarithmic differentiation:

$$\frac{y_1''y_2 - y_2''y_1}{y_1'y_2 - y_2'y_1} - 2\frac{y_2'}{y_2} = \frac{\eta''}{\eta'},$$

or, by virtue of (27):

(29) $$\frac{\eta''}{\eta'} = -p - 2\frac{y_2'}{y_2}.$$

On further differentiation it follows that:

$$\frac{\eta'''}{\eta'} - \left(\frac{\eta''}{\eta'}\right)^2 = -p' - 2\frac{y_2''}{y_2} + 2\left(\frac{y_2'}{y_2}\right)^2,$$

and therefore, by combination with (29):

$$[\eta]_Z = -\frac{1}{2}p^2 - p' - 2\frac{y_2''}{y_2} - 2p \cdot \frac{y_2'}{y_2}.$$

Now the terms which here, on the right side of the equation, contain y_2 are just equal to $2q$ by the differential equation of the second order to which y_2 is subject. *We therefore find:*

(30) $$[\eta]_Z = 2q - \frac{1}{2}p^2 - p',$$

which is the final formula which we sought.

If to every linear differential equation of the second order

(25) there thus belongs a definite differential equation of the third order (26), then clearly to every differential equation (26) belong infinitely many equations (25). We have only to put

$$(31) \qquad 2q - \frac{1}{2}p^2 - p' = r,$$

and in this p (as a rational function of Z, if we lay stress on that point) can still be taken arbitrarily, q being hereupon uniquely determined (and in fact again as a rational function of Z if p and r are rational).

Evidently (26) is completely solved, if one of the corresponding equations is so too. *Conversely, too, the solutions of* (25) *are very readily given if the solutions of the corresponding equation* (26) *are regarded as known.* We conclude, namely, from (27) by integration in the well-known manner:

$$(32) \qquad y_1'y_2 - y_2'y_1 = ke^{-\int pdZ},$$

understanding by k the constant of integration. Combining this with (28), there results:

$$(33) \qquad \begin{cases} y_1 = \eta \cdot y_2, \\ y_2 = \sqrt{\frac{k}{\eta'}} \cdot e^{-\frac{1}{2}\int pdZ}. \end{cases}$$

The linear differential equation of the second order, therefore, requires, after previous solution of the corresponding differential equation of the third order, only a single quadrature besides in order to solve it.

§ 8. Actual Establishment of the Differential Equation of the Third Order for $z[Z]$.

In order now to actually establish the differential equation of the third order:

$$[\eta]_Z = r(Z),$$

which our z satisfies as a particular solution, we make use of what is contained in formula (19) with regard to the development of $(z - z_0)$ in a series according to powers of $(Z - Z_0)$. We consider the developments in series to be explicitly written down, and from them a series calculated for $[z]_Z$ by direct differentiation. As initial term of this series (which, by the way, must proceed according to integral powers of $(Z-Z_0)$ since $[z]_z$ is a rational function of (Z)), we have for $Z_0=1, 0, \infty$, respectively:

$$\frac{\nu_1^2-1}{2\nu_1^2(Z-1)^2}, \quad \frac{\nu_2^2-1}{2\nu_2^2 \cdot Z^2}, \quad \frac{\nu_3^2-1}{2\nu_3^2 \cdot Z^2}.$$

Now I say further, that $[z]_Z$ will certainly not become infinite for a position Z_0 which is different from 1, 0, or ∞. At such a position we have, viz. (as follows again from the conformal representation):

$$z-z_0=a(Z-Z_0)+b(Z-Z_0)^2+\ \ldots$$

where $a \gtrless 0$, and hence for $[z]_Z$ a series proceeding by integral powers of $(Z-Z_0)$ and only possessing positive exponents. We put in accordance with these results:

$$r(Z)=\frac{\nu_1^2-1}{2\nu_1^2(Z-1)^2}+\frac{A}{Z-1}+\frac{\nu_2^2-1}{2\nu_2^2 \cdot Z^2}+\frac{B}{Z}+C,$$

where A, B, C, will be constants, and these we must now so determine, that the development in series, which $r(Z)$ admits in ascending powers of $\frac{1}{Z}$ in the neighbourhood of $Z=\infty$, shall possess the initial term just given $\frac{\nu_3^2-1}{2\nu_3^2 \cdot Z^2}$. *The result shows that A, B, C are completely determined by this necessity.* In fact, we have immediately:

$$C=0,\ A+B=0,\ \frac{\nu_1^2-1}{2\nu_1^2}+\frac{\nu_2^2-1}{2\nu_2^2}+A=\frac{\nu_3^2-1}{2\nu_3^2}.$$

Introducing these, our differential equation will be simply:

$$(34) \qquad [\eta]_Z=\frac{\nu_1^2-1}{2\nu_1^2(Z-1)^2}+\frac{\nu_2^2-1}{2\nu_2^2 \cdot Z^2}+\frac{\frac{1}{\nu_1^2}+\frac{1}{\nu_2^2}-\frac{1}{\nu_3^2}-1}{2\,(Z-1)\,Z},$$

where now for ν_1, ν_2, ν_3, the numerical values of our table (3) may be substituted.*

The three critical points $Z=1, 0, \infty$, just because one of them lies at $Z=\infty$, do not enter into this differential equation

* For the binomial equation (1) we get as the corresponding differential equation by direct differentiation:

$$[\eta]_Z=\frac{n^2-1}{2n^2}\cdot\frac{1}{Z^2}.$$

with a symmetry corresponding to their peculiar importance. We shall at once remedy this if we introduce in place of Z as a new variable some linear function of Z, which for $Z=1, 0, \infty$, assumes any three finite values a_1, a_2, a_3. Making use of the formula (24), but, be it noted, calling the new variable itself Z again, we have:

$$(35)\quad [\eta]_Z = \frac{1}{Z-a_1 \,.\, Z-a_2 \,.\, Z-a_3}\left\{\frac{\nu_1^2-1}{2\nu_1^2(Z-a_1)}(a_1-a_2)(a_1-a_3) + \frac{\nu_2^2-1}{2\nu_2^2(Z-a_2)}(a_2-a_3)(a_2-a_1) + \frac{\nu_3^2-1}{2\nu_3^2(Z-a_3)}(a_3-a_1)(a_3-a_2)\right\},$$

where now, as we see, all desirable symmetry exists.

§ 9. Linear Differential Equations of the Second Order for z_1 and z_2.

The developments of § 7 put us in a position to give the most general linear differential equation of the second order with rational coefficients:

$$(36)\qquad y''+p\cdot y'+q\cdot y=0,$$

which has two particular solutions y_1, y_2, whose quotient is equal to our z; we have only to put, according to formulæ (31), (34):

$$2q-\frac{1}{2}p^2-p' = \frac{\nu_1^2-1}{2\nu_1^2(Z-1)^2}+\frac{\nu_2^2-1}{2\nu_2^2\,.\,Z^2}+\frac{\frac{1}{\nu_1^2}+\frac{1}{\nu_2^2}-\frac{1}{\nu_3^2}-1}{2(Z-1)Z}$$

I say now *that among these differential equations there is always one which the roots z_1, z_2 of our form-problem satisfy.* In fact, we recognise *a priori* that z_1, z_2 must be particular solutions of a linear differential equation of the second order with rational coefficients. Namely, let z_1^0, z_2^0, be two corresponding branches of our functions, then any other branches express themselves as linear homogeneous functions of these z_1^0, z_2^0. They therefore all satisfy the following differential equation:

$$\begin{vmatrix} y'' & y' & y \\ \dfrac{d^2z_1^0}{dZ^2} & \dfrac{dz_1^0}{dZ} & z_1^0 \\ \dfrac{d^2z_2^0}{dZ^2} & \dfrac{dz_2^0}{dZ} & z_2^0 \end{vmatrix} = 0.$$

We now conclude at once that the coefficients, which y'', y', y obtain when this determinant is developed, behave as rational functions of Z. They are themselves, indeed, without further consideration, rational functions. For if we replace z_1^0, z_2^0 by any other pair of corresponding branches of z_1, z_2:

$$\alpha z_1^0+\beta z_2^0,\ \gamma z_1^0+\delta z_2^0,$$

these coefficients, since $\alpha\delta-\beta\gamma$ by virtue of the definition of the form-problem $=1$, remain altogether unaltered, according to the rule for the multiplication of determinants. Our object now is to seek, out of the totality of the differential equations (36), the one which z_1 and z_2 satisfies.

Let y_1, y_2 be two solutions of (36), such that $\frac{y_1}{y_2}=z$. *Then we will first calculate generally:*

$$X(y_1, y_2)=\frac{F_2(y_1, y_2)\,.\,F_3(y_1, y_2)}{F_1(y_1, y_2)}$$

To this end we start from the equation

$$c\,.\,\frac{F_2^{\nu_2}(z, 1)}{F_3^{\nu_3}(z, 1)}=Z.$$

Differentiating this, and considering as before that F_1 is always, save as to a numerical factor, the functional determinant of F_2 and F_3, we obtain (c' representing a proper constant):

$$c'\,.\,\frac{F_2^{\nu_2-1}(z, 1)\,.\,F_1(z, 1)}{F_3^{\nu_3+1}(z, 1)}\,.\,z'=1,$$

or, on introducing another appropriate multiplier c'':

$$c''\,.\,Z\,.\,\frac{F_1(z, 1)}{F_2(z, 1)\,.\,F_3(z, 1)}\,.\,z'=1.$$

Here let us now put $z=\frac{y_1}{y_2}$. Then

$$c''\,.\,Z\,.\,\frac{F_1(y_1, y_2)}{F_2(y_1, y_2)\,.\,F_3(y_1, y_2)}\,.\,(y_1'y_2-y_2'y_1)=1,$$

or finally, embodying the symbol X and the formula (32) also:

$$(37)\qquad X(y_1, y_2)=k\,.\,c''\,.\,Z\,.\,e^{-\int p\,dz},$$

which is the formula we required.

Now for the solutions z_1, z_2 of our form-problem, not only was $\frac{z_1}{z_2}=z$, but it was determined that $X(z_1, z_2)$ was to be independent of Z. *We shall therefore have to take the coefficients p of the corresponding linear differential equation in such a way that Z disappears altogether from the corresponding formula* (37). This gives, as we see,

$$e^{\int p dZ}=Z \text{ or } p=\frac{1}{Z}.$$

Introducing this value into (36), we obtain the differential equation which we sought. This, after some easy modifications, runs as follows: *

$$(38)\; y''+\frac{y'}{Z}+\frac{y}{4(Z-1)^2 \cdot Z^2}\cdot\left\{-\frac{1}{\nu_2^2}+Z\left(\frac{1}{\nu_2^2}+\frac{1}{\nu_3^2}-\frac{1}{\nu_1^2}+1\right)-\frac{Z^2}{\nu_3^2}\right\}=0.$$

§ 10. Relations to Riemann's *P*-Function.

We now have all we require in order to calculate by a series of powers z_1, z_2, and from them $z=\frac{z_1}{z_2}$, in the neighbourhood of any position $Z=Z_0$. In fact, we saw in § 5 how we could determine in an individual case the nature of this series of powers, and have now simply to substitute the series itself in (38) in order to find the coefficients in the series which still remain unknown. If we wish to effect this in particular for the neighbourhood of the point $Z=\infty$, we can use the formulæ (21) immediately.

If I do not more explicitly carry out the step here proposed, nor discuss more closely the convergence and the analytical law of progression of the developments suggested, it is because we have meanwhile obtained all the preliminary conditions for basing the investigation of the functions z_1, z_2 on a ready-prepared and well-known theory. *I mean the theory of Riemann's P-functions:*

$$P\begin{pmatrix}\alpha & \beta & \gamma & \\ \alpha' & \beta' & \gamma' & x\end{pmatrix}$$

* For the solutions z_1, z_2 of the form-problem of the cyclic group, we find in a similar way:

$$y''+\frac{y'}{Z}-\frac{y}{4n^2Z^2}=0.$$

*and the representation of their several branches by the hypergeometrical series of Gauss.** I have already said that I will not take for granted any previous special knowledge concerning the P-functions. We may therefore define these functions in the way which most conveniently fits in with our previous developments, *viz.*, *as solutions of the following differential equation of the second order:*

$$(39)\qquad P''+\frac{P'}{x(1-x)}[(1-a-a')-(1+\beta+\beta')\,x]$$
$$+\frac{P}{x^2(1-x)^2}[aa'-(aa'+\beta\beta'-\gamma\gamma')\,x+\beta\beta'x^2]=0,$$

where $a+a'+\beta+\beta'+\gamma+\gamma'$ is always to be taken equal to 1.† Clearly (38) is a special case of (39); to obtain (38) we have only to write:

$$P=y,\ x=Z,\ a=-a'=\frac{1}{2\nu_2},\ \beta=-\beta'=\frac{1}{2\nu_3},\ \gamma=\frac{1}{2\nu_1},\ \gamma'=\frac{\nu_1^2-1}{2\nu_1},$$

which is reconcilable with the condition $a+a'+\beta+\beta'+\gamma+\gamma'=1$, since ν_1 is in all our cases $=2$. *Therefore* z_1, z_2 *are with reference to the particular value of* ν_1, *special cases of the function:*

$$(40)\qquad P\begin{pmatrix} \frac{1}{2\nu_2} & \frac{1}{2\nu_3} & \frac{1}{4} & \\ & & & Z \\ -\frac{1}{2\nu_2} & -\frac{1}{2\nu_3} & \frac{3}{4} & \end{pmatrix}.$$

We can now characterise more precisely our functions z_1, z_2 among the general ones denoted by this symbol. It is just for this purpose that I have established the formulæ (21) explicitly. If in these we multiply z_1 by $Z^{\frac{1}{2\nu_3}}$ and z_2 by $Z^{-\frac{1}{2\nu_3}}$, the products remain finite for $Z=\infty$, and different from zero, and, moreover,

* Any one who wishes to enter on these theories will find it still the best plan, in addition to *Gauss's* "Disquisitiones generales circa seriem infinitam," &c. (1812, Works, t. iii.), and *Kummer's* memoirs on the hypergeometrical series (1836, Crelle's Journal, Bd. 15), to study the original work of *Riemann:* "Beiträge zur Theorie der durch die Gauss'sche Reihe $F(a,\beta,\gamma,x)$ darstellbaren Functionen" Bd. 7 der Göttinger Abhandlungen (1857), or Werke, p. 62–82).

† This differential equation is obtained by an easy modification from that which Riemann gives specially for $P\begin{pmatrix} a & \beta & 0 & \\ a' & \beta' & \gamma' & x \end{pmatrix}$ (Werke, p. 75).

are continuous in the neighbourhood of the point $Z = \infty$. *The formulæ* (21) *there denote just such series as Riemann introduced l. c. under the title* $P^{(\beta)}$, $P^{(\beta')}$; only that Riemann leaves undetermined the first coefficients of $P^{(\beta)}$ and $P^{(\beta')}$. If we choose them, in particular, as is done in the formulæ (21), we can say, finally, *that our* z_1, z_2 *are specially those among the general P-functions* (40), *which spring from the series* $P^{(\beta)}$, $P^{(\beta')}$, *by any analytical expansion.*

With this theorem we have reached the object of the developments of the present chapter. I wished to show that our functions z, z_1, z_2 belong to those into which the modern theory of functions, both by its geometrical representations and its analytical weapons, obtains a, so to say, *complete* insight. Granted this, we have thus at the same time attained to a point of view which is to serve us in the second part of our exposition, viz., it then appears reasonable to reduce more complicated algebraical functions, so far as is possible, to our present ones z, z_1, z_2.

But, moreover, the developments here given can only be considered, even more so than our other ones, as an *introduction.* In fact, our intention of putting the argument in the most elementary form possible has hindered us from explaining a point which is really the most interesting, viz., how the linear substitutions to which we have subjected z, z_1, and z_2 respectively in the preceding chapter now come into prominence, when we look upon z, z_1, z_2 as functions of Z, and allow the latter variable to traverse a closed path in its plane. We should also have been able, if we had followed the proposition given in § 5 a little further, to find the direct transition to Riemann's P-function without previously having formulated explicitly the differential equations. I leave it to the reader to familiarise himself, by his own studies and reflections, with these and allied questions.

CHAPTER IV

ON THE ALGEBRAICAL CHARACTER OF OUR FUNDAMENTAL PROBLEM

§ 1. Problem of the Present Chapter.

Having in the previous chapter discussed our fundamental problems only under the aspect of the theory of functions, let us now treat them from the point of view of the theory of equations. I understand by this latter, the aggregate of the theories which relate to the *rational resolvents, i.e.*, to those auxiliary equations which any rational functions of the roots of the given equation satisfy.

A first and important portion of this theory, which distinguishes the *nature* of the resolvents coming generally under consideration, is formed by those reflections which, in accordance with the fundamental ideas of Galois, are usually denoted by his name, and which amount *to characterising the individual equation, or system of equations, by a certain group of interchanges of the corresponding solutions* (the word *group* being taken in the same specific sense which we have explained in the first chapter). I will, in paragraphs 2–4 following, make mention of the foundations of this theory so far as seems necessary for understanding what follows, but I refer otherwise to the text-books already mentioned above,* and this not only for the more thorough completion, *but especially for the proofs.* On this basis it is very easy to characterise our fundamental problems in Galois's sense (§§ 5, 6). In particular, it follows that these must all admit of solution by extraction of roots, with the sole exception of the icosahedral equation, whose lowest resolvents are of the fifth and sixth degrees respectively. I

* See remark on p. 6 *supra.*

shall, in the concluding remarks of this chapter (§ 16), draw attention more in detail to the prime importance of this result.

Howbeit, it is not sufficient in any given algebraical problem to know the *nature* of the resolvents; we require, further, *to actually calculate these resolvents, and this in the simplest manner.* The second part of the present chapter is concerned with this, with strict limitation to the questions immediately surrounding our fundamental problems. I show, first of all (§ 7), how we can actually construct the auxiliary resolvents by means of which the solution of the dihedral, tetrahedral, and octahedral equation is to be achieved. I concern myself, then, in detail with the resolvents of the fifth and sixth degrees of the icosahedral equation (§§ 8–15). The particular equations of the fifth and sixth degrees, which we so obtain, will be of essential importance for our later developments. Here it is primarily the *method* on which I wish to lay stress now; a method which makes use at one time of the theory of functions, at another of the theory of invariants, and in both directions seems capable of an extension to higher problems.

§ 2. On the Group of an Algebraical Equation.

Our object now being to define the *group* which belongs to each individual algebraical equation from the point of view of Galois's theory, we will first consider the classification which we can derive for the rational functions of n variable magnitudes:

$$x_0, x_1, \ldots\ldots x_{n-1},$$

from their behaviour towards the permutations of the x's. It is clear *a priori* that all permutations of the x's which leave unaltered such a rational function form a group which is contained as a sub-group in the totality of the permutations (or, perhaps, is identical with this totality). But the converse is also the case; as soon as any group of permutations of the x's is given, we can always construct such rational functions of the x's as remain unaltered for the permutations of this group, but for no other. We call these rational functions of the x's *those belonging to the group of permutations*, and now classify generally all rational functions of the x's which occur according to the group of permutations to which they belong.

We must further acquaint ourselves with the so-called *theorem of Lagrange.** Let R and R_1 be two rational functions of the x's, and let R remain unaltered by all permutations which make up the group appertaining to R_1 (where, of course, it is not stated that R must belong to the same group). Let, further, $s_1, s_2, \ldots s_n$ be the elementary sums of powers:

$$(1) \qquad s_1 = \sum x, \; s_2 = \sum x^2, \; \ldots \; s_n = \sum x^n.$$

Then the theorem alluded to declares that R can be represented as a rational function of R_1 and $s_1, s_2, \ldots s_n$. We can easily generalise this theorem still further by considering, instead of R_1, a number of rational functions: $R_1, R_2, \ldots$ to be given, and assuming that R remains unchanged by all those permutations which leave $R_1, R_2, \ldots$ *simultaneously* unaltered. *Then R will be a rational function of $R_1, R_2, \ldots$ and the $s_1, s_2, \ldots s_n$.* In fact, we can compose rationally of the $R_1, R_2, \ldots$ a rational function R' of the x's which only remains unaltered for those permutations of the x's which leave $R_1, R_2, \ldots$ simultaneously unaltered. According to the first application which we made of the theorem of Lagrange, R will then be capable of being represented rationally by means of this R' and the $s_1, s_2, \ldots s_n$, whereupon our new assertion is proved *eo ipso*.

Now let the equation of nth degree be given:

$$f(x) = 0,$$

whose roots are to be the $x_0, x_1, \ldots x_{n-1}$ previously considered. Then, in any case, we know the values of the s_i (1); and hence, by rational processes of operation, the rational symmetric functions of the x's generally. But it may happen that some unsymmetric functions of the x's : $R_1, R_2, \ldots$ are given us. Then we can, on the ground of the expanded Lagrange theorem, compute generally every function R of the x's in a rational manner, which remains unaltered for all permutations which at the same time leave $R_1, R_2, \ldots$ unaltered. *Therefore we shall always have those rational functions of the x's, and only those, "rationally known" (as we will say), which remain unaltered for a determinate group of permutations of the x's.*

* "Réflexions sur la résolution algébrique des équations." Mem. de l'Acad. de Berlin, t. iii. (1770–71), or Œuvres, t. iii. (§ 100 of the Memoir).

The theory here sketched is first applicable, as we said, to the case of x's altogether independent. *But now the point is that there exists in every special case also an analogous theory.* If, in such a case, we say of a function that it remains unaltered for certain permutations, we understand thereby that it does not change its *numerical* value. *There is always, then, such a group G of permutations of the x's that all rational functions of the x's which remain unaltered for G, and only these, are rationally known. Besides this, the law holds good that all permutations of G which leave unaltered any given rational function of the x's in each case form a group, so that, in relation to the permutations of G, the classification of rational functions just described and also the theorem of Lagrange are retained with no exception.* The group G is then that which *Galois* describes as the *group of the equation.**

The difficulties of the Galois theory lie, perhaps, less in the general theorems here formulated than in the notion of being "rationally known" which is employed in them. When shall we apply this description to functions? We *must* do so if (in consequence of special values of $x_0, x_1, \ldots$) they have rational values, *i.e.*, are equal to rational functions of the s_i (with rational numerical coefficients). But we *can* do so for quite arbitrary functions $R_1, R_2, \ldots$ if we assume that we have already by some means computed the values of $R_1, R_2, \ldots$ We then *adjoin*, as *Galois* expresses it, these $R_1, R_2, \ldots$ and accordingly widen the *rationality domain*, to use the language of *Herr Kronecker*,† in which we operate. In this sense the statements which the Galois theory makes concerning the individual equation $f(x)=0$ are to a certain degree dependent on our subjective interpretation. If we adjoin the whole of the roots of $f(x)=0$, the group of the equation always consists of identity alone. We must therefore abandon the conception that an equation of the nth degree with a group which we describe as of limited extent must therefore necessarily have in any sense specified coefficients.

* See "Œuvres de Galois," in Liouville's Journal, t. xi. (1846).

† *Cf.* here, Kronecker, "Grundzüge einer arithmetischen Theorie der algebraichen Grössen" (Bd. 92 of the Journal für Mathematik, 1881).

§ 3. General Remarks on Resolvents.

Now let G again be the group of the equation $f(x)=0$, N the order of the group. The only assumption to which we subject G is that of being *transitive*, *i.e.*, of embracing permutations in virtue of which the individual root x_k of $f=0$ can replace any other root x_l. Otherwise $f(x)=0$ would be *reducible*, *i.e.*, would split up into rational factors, and we should therefore be able, instead of $f(x)=0$, more effectively to consider the several equations which arise from equating to zero the individual factors.

We now choose any rational function R_0 of the roots x, such as does not remain unaltered for all the permutations of G, and therefore is not rationally known, though it may remain unaltered for some permutations in number ν, which form a group g_0. For the permutations of G, R_0 assumes on the whole $\frac{N}{\nu}=n'$ different values:

$$R_0, R_1, \ldots\ldots R_{n'-1}.$$

We then form the equation on which these different values depend:

$$(R-R_0)(R-R_1)\ldots\ldots(R-R_{n'-1})=0.$$

We have thus evidently obtained an equation whose coefficients are rationally known, for they are symmetric functions of the different R's, and, as such, invariant for the permutations of G. This is what we denote as a *resolvent* of the foregoing equation $f(x)=0$, and indeed, when this may be of importance, as a *rational* resolvent, inasmuch as on it a rational function of the x's depends.

We inquire as to the totality of the different kinds of resolvents which $f(x)=0$ possesses. In this respect we may make the following convention beforehand. If we had chosen instead of R_0 another rational function of the roots, which equally appertains to g_0, it would, by Lagrange's theorem, admit of rational expression in terms of R_0 and the known rational quantities; *the new resolvent would therefore result from the former (and, similarly, the former from the new one) by rational transformation.* We will agree to look upon as altogether identical two resolvents of this kind in the general survey of them which will be given here. Then to every group g_0 there appertains always only one corresponding resolvent.

But the same resolvent also arises if we start from certain other sub-groups instead of g_0. In fact, instead of beginning with the root R_0, we can, in the construction of the resolvent, just as well put one of the other roots $R_1, R_2, \ldots$ in the foreground. Then, in place of g_0 those groups of permutations of the x's occur which respectively leave unaltered $R_1, R_2, \ldots$ and which we will denote by $g_1, g_2, \ldots$. We inquire how these g_i's are connected with the original g_0. Let S_i be one of those permutations of the x's by which R_i is transformed into R_0; the totality of such permutations will then be given by $S_i T^{(0)}$, understanding by $T^{(0)}$ the several permutations of g_0 in turn. We now combine with $S_i T^{(0)}$ the inverse operation S_i^{-1}. Then R_0 is transformed back into R_i. *Hence* R_i *remains unaltered for all permutations:*

$$T^{(i)} = S_i \, . \, T^0 \, . \, S_i^{-1}.$$

Now, conversely, from every $T^{(i)}$ for which R_i remains unaltered, a $T^{(0)}$ can be derived by the corresponding method in the form:

$$T^0 = S_i^{-1} \, . \, T^{(i)} \, . \, S_i.$$

This new formula is, as we see, the immediate solution of that just given; we have therefore in this latter defined the whole of the permutations generally which leave R_i unaltered, *i.e.*, the group g_i. *The group* g_i *therefore proceeds from* g_0 *through transformation by* S_i.

Now S_i (if we take into consideration all the roots $R_0, R_1, \ldots R_{n-1}$) can here be any arbitrary permutation of G. For by S_i^{-1} some one of the R_i must always proceed from R_0. Consequently, we can describe the groups $g_1, g_2, \ldots g_{n-1}$ as the totality of those which exist within G by transformation from g_0. Such groups we have previously described as associates. Hence we have, finally, to sum up what has gone before, the concise theorem: that *there are as many different kinds of resolvents of a proposed equation* $f(x) = 0$ *as there exist different systems of associate sub-groups within the corresponding group* G.

We now determine the group Γ of the individual resolvent so obtained. I say that *it will be constructed of those permutations of the* R's *which occur when we subject the* x's *to the permutations of* G. For a rational function of the R's which remains unaltered under the said permutations of the R's is, at the same time, when considered as a function of the x's, unchanged for the permutations of G, and, conversely, it cannot be the latter if the former is not the case. *The group* Γ *is, therefore, for every case, isomorphic with the group* G.

Here we must now make an important distinction. The isomorphism which has been found can be simple or multiple. The latter occurs then, and only then, when such permutations of the x's exist within G, as leave unaltered *the whole of the* R_i's; these permutations will then form a group γ, which is self-conjugate within G. The resolvent plays an entirely different part with respect to the original equation in the two cases.

In the first case, we can compose rationally every rational function of the x's, and in particular the x's themselves, from the R_i's, with the help of the known quantities. The original equation is, therefore, itself a resolvent of the resolvent: the solution of the one equation ensures that of the other, and conversely. On replacing the equation $f(x)=0$ by its resolvent, it is true we have attained a modification of the original problem, but in no way a simplification thereof.

It is quite otherwise in the second case. The x's are in it by no means rational in the R_i's. If we have computed the R_i's, the original equation $f(x)=0$ has yet to be solved. This problem is now simplified only so far as the group G is now (after adjunction of the R_i's) replaced by γ.* But, on the other hand, the determination of the R_i's themselves is more easy to carry out than the computation of the x's: *for the group* Γ *of the corresponding equation is smaller than* G. We have therefore decomposed the original problem into two steps of a more simple character.

Clearly the resolvents of the second kind are the more important. They can only occur when the group G of the proposed equation is *compound*. By studying in such a case the decomposition of G, we have, at the same time, the means of simplifying, step by step, the equation $f(x)=0$, by means of a complete series of resolvent auxiliary equations. It is just this significance of resolvents which the ordinary theory makes use of in the solution of the equations of the third and fourth degrees.

* Hereby $f(x)=0$ may possibly have become reducible (even if γ, when expressed in terms of the x's, is not transitive).

§ 4. The Galois Resolvent in Particular.

According to what has just been said, all resolvents whose group Γ is simply isomorphic with the group G of the proposed equation $f(x)=0$ represent, in the abstract, equivalent problems. But there is one amongst them which, for the purpose of algebraical exposition, possesses quite a special significance: *it is that which we are accustomed to call by the name of the Galois resolvent, and which is defined by the fact that its individual roots are altered for every permutation of the x's which is contained in G.* Therefore the groups $g_0, g_1, \ldots$, which we just now made to correspond to $R_0, R_1, \ldots$, then all reduce to identity, and simultaneously the degree of the resolvent becomes as high as possible, viz., equal to N. On the other hand, it offers this advantage, that we need only compute *one* of its roots. In fact, by Lagrange's theorem *all* rational functions of the x's must express themselves rationally in terms of this one root and the known quantities.

But let us consider more closely the properties of the *Galois* resolvent.

First as regards its group; for every one of the N-operations of the group G, each of the N-roots

$$R_0, R_1, \ldots N_{N-1}$$

will be replaced. There are, therefore, no two operations of G which would both bring the same root R_i into the same position R_k: the individual operation is fully determined provided only we know in what way it influences an individual R_i. Introducing the notion of transitivity, as it has already been used, we can say:

The group Γ of the Galois resolvent is just simply transitive.

We can, therefore, denote the individual permutation of Γ by the index of that root R_k which proceeds from R_0 by means of it. In this sense we will forthwith make use of the symbol S_k.

We now express rationally, by means of the theorem of Lagrange, the different roots $R_0, R_1, \ldots R_{N-1}$, in terms of the first of them. In this manner N formulæ arise, which we write in the following way:

$$(2) \qquad R_0=\psi_0(R_0), R_1=\psi_1(R_0) \ldots R_{N-1}=\psi_{N-1}(R_0).$$

Here the ψ_i's denote rational functions of the accompanying argument, which are only so far completely determinate that we shall not modify them by the help of the Galois resolvent itself, and $\psi_0(R_0)$ is of course only written instead of R_0 itself for the sake of uniformity. We select one of these formulæ and write (neglecting the former indices of the R's):

$$(3) \qquad R' = \psi_i(R),$$

and consider the Galois resolvent transformed by the help of this formula (by eliminating the R between the resolvent and the formula (3)). Thus arises an equation of the order N for R' which, in any case, has the root R_i in common with the original Galois resolvent. Now, the resolvent is by hypothesis irreducible. Hence the two equations of the Nth degree have all their roots common, *i.e.*, they are identical. We have, therefore, the theorem:

The Galois resolvent will be transformed into itself by the N rational transformations (3).

If we therefore substitute in formula (3), instead of R, any root R_k, R' will become equal to another root R_j. But, instead of R_k, we can write $\psi_k(R_0)$, and $\psi_j(R_0)$ instead of R_j. Hence:

$$\psi_j(R_0) = \psi_i\psi_k(R_0),$$

and therefore generally:

$$\psi_j = \psi_i\psi_k,$$

so far, namely, as we disregard the changes which can be wrought on the individual symbols of this expression by the help of the Galois equation satisfied by the R_i. In this sense we have:

The N rational transformations (3) *form a group.*

We ask how this group is connected with the Galois group Γ. If we replace, in the formulæ (2), the R_0 on the right hand by $R_0, R_1, \ldots R_{N-1}$ in order, we obtain on the left-hand side, in consequence of what has just been said, the roots R_i again, in each case in altered sequence. We obtain, therefore, N different arrangements of the R's, and now the assertion may be proved that *those N permutations, by which these arrangements proceed from the original arrangement, just make up the group* Γ. For this purpose we will show that a rational function of the R_i's

$$F(R_0, R_1, \ldots R_{N-1}),$$

which remains unaltered when we replace the sequence R_0, $R_1, \ldots R_{N-1}$ by any of the other N orders in question, is rationally known. In fact, every rational function of the R_i's can in virtue of (2) be compressed into the form $\Phi(R_0)$. If, now, F admits the changes mentioned, it will be just as truly equal to $\Phi(R_1)$, or equal to $\Phi(R_2)$, &c., understanding in every case by Φ the same rational function. Therefore also

$$F = \frac{1}{N}[\Phi(R_0) + \Phi(R_1) + \ldots \Phi(R_{N-1})];$$

therefore F is equal to a symmetric function, and hence, in fact, can be rationally computed, as was asserted.

The relation between Γ and the group of the transformations (3) thus found we will investigate more closely. If we put R_k instead of R_0 on the right-hand side of (2), R_k appears also on the left-hand side in the first position. We therefore obtain the same order of the R_i's as proceeds from the original one by the operation S_k of Γ. Now writing instead of R_k (on the right-hand side) $\psi_k(R_0)$ throughout, we can say as follows:

The operation S_k is that which replaces $\psi_i(R_0)$ $(i = 0, 1, \ldots (N-1))$ by $\psi_i\psi_k(R_0)$.

Similarly the operation S_l will be that which replaces $\psi_i(R_0)$ by $\psi_i\psi_l(R_0)$, or, what is the same thing, which replaces $\psi_i\psi_k(R_0)$ by $\psi_i\psi_k\psi_l(R_0)$ (where in both places we will allow i to range from 0, 1, to $(N-1)$). If we combine the two theorems thus obtained, by applying first S_k and then S_l, it follows that:

For the operation S_kS_l, $\psi_i(R_0)$ will be replaced by $\psi_i\psi_k\psi_l(R_0)$.

The relation which we find in this form between the groups of the S's and of the ψ's is at first not one of isomorphism. For S_kS_l denotes that we first apply S_k and then S_l, while $\psi_k\psi_l(R_0)$ says that we first compute the ψ_l of R_0 and from it the ψ_k. But we can directly so modify the relation that isomorphism results. To this end we need only make S_k to correspond to the *inverse* operation ψ_k^{-1}. In fact $(\psi_k\psi_l)^{-1} = \psi_l^{-1} \cdot \psi_k^{-1}$. Hence we have:

The groups of the S's and of the ψ's are simply isomorphic.

The theorems thus formulated are the more important because we can reverse them without further trouble. In fact, we find, if we repeat what has been already said in a different order:

If an irreducible equation of the Nth degree is transformed into itself by N rational transformations:

$$R' = \psi_0(R),\ R' = \psi_1(R), \ldots\ldots,$$

*it is its own Galois resolvent, and its group Γ stands to the group of the ψ's in the relation just explained.**

If, then, for such an equation, a rational function of the roots is constructed which remains unaltered for the permutation S_k of a certain sub-group contained in the Galois group, and thus can be introduced as a root of a corresponding resolvent, it is sufficient to establish a rational function of the single root R_0, which will, for the corresponding ψ_k's, be transformed into itself; for the sub-group of the ψ_k's contains at the same time all the (ψ_k^{-1})'s, and, therefore, corresponds to the sub-group of the S_k's in the isomorphic co-ordination.

§ 5. Marshalling of our Fundamental Equations.

I have framed the foregoing paragraph in such detail in order to be able to now marshal directly our fundamental equations in the scheme of the Galois theory, to wit, the *binomial* equations and the equations of the *dihedron, tetrahedron, octahedron, and icosahedron.* Let us first agree that our equations are irreducible. From the considerations in the last chapter, based on the function theory, it follows, that the N-function branches, which are defined by the individual equations, on regarding in each case the right-side Z as independent variable, are all connected with one another. *Therefore the hypotheses are exactly fulfilled to which the concluding theorem of the preceding paragraph relates.* For the N-roots which any particular one of our equations possesses do in fact proceed

* This theorem must not be confused (as it occasionally has been) with the definition of the Abelian equations. For these also there are N rational transformations $R' = \psi_i(R)$, but it is further assumed that the ψ's are permutable, and that therefore $\psi_i\psi_k = \psi_k\psi_i$.

from any one of their number in each case by N rational transformations, viz., by the N *linear substitutions* well known to us.

Thus we have at once: *our equations are their own Galois resolvents,* and we can now immediately draw further conclusions by adopting what was said above concerning the groups of the corresponding (non-homogeneous) linear substitutions.

Let us first select, say, the *octahedron,* and recall that the group was composed of the 24 octahedral substitutions. In it the tetrahedral group of 12 substitutions was contained as the most comprehensive self-conjugate sub-group; in this again the quadratic group (of 4 substitutions), and in the latter, finally, a cyclic group of 2 substitutions. We conclude therefore: *that we can solve the octahedral equation by a series of 4 auxiliary equations whose groups are respectively* $\frac{24}{12}, \frac{12}{4}, \frac{4}{2}, 2,$ *i.e.,* contain 2, 3, 2, 2 *permutations.* A group whose degree is a prime number is necessarily a cyclic group. If now with Lagrange we add to this that every cyclic equation of the nth degree can be replaced by a binomial equation of the nth degree,* we recognise that: *the octahedral equation can be solved by extracting in succession a square root, then a cube root, and, finally, two more square roots.* We will confirm this in § 7 by explicit formulæ.

As regards the *tetrahedral equation,* this is itself solved at the same time by what was said concerning the octahedral equation; for the tetrahedral group is a self-conjugate subgroup of the octahedral group. For the *dihedral equation of degree* $2n$, we find that it must admit of reduction to a binomial equation of the nth degree by extraction of a square root. And finally, the solution of the *binomial equation* itself can then, and only then, be decomposed into several steps when its degree is a composite number.

Thus the *icosahedral equation* stands alone by the side of the binomial equation of prime degree, as the only one of our equations which we cannot reduce by the construction of resolvents.

* The equation of the nth degree is called *cyclic* if its Galois group is cyclic, and therefore contains, say, only the cyclic permutations of $(x_0, x_1, \ldots x_{n-1})$. The method then consists, as is well known, in introducing as the unknown magnitude $x_0 + \epsilon x_1 \ldots \epsilon^{n-1} x_{n-1}$, where $\epsilon = e^{\frac{2i\pi}{n}}$.

If we wish to construct resolvents for it as well (as we do in § 8 following), our earlier researches on the icosahedral group teach us that, as the lowest resolvents, those of the fifth and sixth degrees come under consideration. The former correspond to the circumstance that the icosahedral group contains 5 associate tetrahedral groups; the latter to the other circumstance that it contains 6 associate dihedral groups of 10 operations each. These resolvents will in both cases possess again a Galois group of 60 permutations. We can say directly, from what has gone before, that these, for the resolvents of the fifth degree, are the 60 even permutations of the roots, and that, therefore, the product of the differences of the roots must be rational. We shall not determine more exactly the group of the resolvent of the sixth degree till later on (§ 15).

While we thus take advantage of the results of our previous investigations for dealing with the Galois theory, we must certainly not overlook one important circumstance. We are only entitled to reckon the linear functions of our substitution groups among the rational functions ψ of the preceding paragraph, provided that we suppose the coefficients occurring in the formulæ of the linear substitutions as rationally known. These are certain roots of unity. *We must therefore suppose these roots of unity adjoined, in order that the foregoing statements may be accurate.* In the case of the icosahedral equation, for instance, we must adjoin the fifth roots of unity, *i.e.*, the numerical irrationalities which are determined by the equation:

$$\frac{x^5-1}{x-1}=0.$$

Let us explain by this example the consequences which would otherwise ensue. It is known that the foregoing equation of the fourth degree has a cyclic group of 4 permutations,* a group therefore which contains a self-conjugate sub-group of 2 permutations. We conclude that the icosahedral equation now possesses a group of 4·60 permutations among which a sub-group of 2·60 permutations, and then one of 60 permutations, is self-conjugate. This new group of the icosahedral equation need by no means necessarily be transferred unchanged to the individual resolvent of the icosahedral equation. Indeed, for

* See, *e.g.*, Bachmann, "Die Lehre von der Kreistheilung," Leipzig (1872).

the resolvents of the fifth degree this is *a priori* not possible, since their group can never contain more than $\underline{|5} = 2\cdot 60$ permutations. In fact, only $\sqrt{5}$ occurs as a numerical irrationality in the formulæ which we shall establish in § 14 for the difference product of our resolvent of the fifth degree, so that the adjunction of the individual fifth root of unity is by no means necessary to reduce the group of the resolvent to only 60 permutations. We do not pursue this matter further, because it would involve us too deeply in considerations appertaining to the theory of numbers.*

§ 6. Consideration of the Form-Problems.

We further consider in a few words the form-problems which run parallel with our equations. These are systems of equations with, in every case, two unknowns, z_1, z_2. We shall be able to apply the fundamental ideas of the Galois theory throughout to these systems of equations by substituting, whenever in these latter mention is made of the roots of an equation, the individual pairs of solutions z_1, z_2. In particular, we shall then be able to say that *our form-problems are their own Galois resolvents.* In fact, all the $2N$ systems of solution which our form-problems possess are derived from the individual systems of solution by $2N$ linear homogeneous substitutions, which are known *a priori.*† It is here, therefore, the *homogeneous* linear substitution-groups of our earlier exposition which determine the Galois group of the problem in question.

These homogeneous groups were all compound, inasmuch as they contained a self-conjugate sub-group, which consisted of identity and the following operation:

$$z_1' = -z_1, \; z_2' = -z_2.$$

We thence conclude that our form-problems must always admit

* We have, in the text, represented the Galois theory as practically known, and then deduced from it properties of the icosahedral equation, &c. On the other hand, the beginner cannot be too strongly recommended to reverse the whole method of consideration, and to employ the properties of the icosahedral equation, &c., in order to extract from them, as a simple example, the general ideas of the Galois theory.

† The roots of unity here occurring figure in the text again as adjoined quantities.

of solution if we first solve an equation with a group of N permutations, and then extract a square root. This is now just what we have already effected in § 2 of the preceding chapter, while dealing with the reduction of the form-problems. It will be superfluous to spend further time over the details of this.

§ 7. The Solution of the Equations of the Dihedron, Tetrahedron, and Octahedron.

Turning now to the communication of the proposed formulæ of solution for the dihedron, tetrahedron, and octahedron, we again commence with the consideration of the *octahedral equation.* We write it as before:

$$(4) \qquad \frac{W^3}{108t^4} = Z.$$

We will then introduce, as a root of the first auxiliary equation, such a rational function of z as remains unaltered for the 12 tetrahedral substitutions. It is clearly the simplest plan to choose for this the right side of the corresponding *tetrahedral equation.* Denoting this by Z_1, we have:

$$(5) \qquad \frac{\Phi^3}{\Psi^3} = Z_1.$$

We further choose, as the unknown of the second auxiliary equation, corresponding to the *quadratic group,*

$$(6) \qquad -\frac{(z_1^2 - z_2^2)^2}{4z_1^2 z_2^2} = Z_2,$$

and, finally, as the unknown of the third auxiliary equation, the right side of the *binomial* formula:

$$(7) \qquad \left(\frac{z_1}{z_2}\right)^2 = Z_3.$$

The fourth auxiliary equation will then simply arise in calculating from this Z_3 the $\frac{z_1}{z_2} = z$ itself.

In order now to actually construct the auxiliary equations on which Z_1, Z_2, Z_3, and, finally, $\frac{z_1}{z_2}$ depend, we need only recall that all rational functions of z, which remain unaltered for the tetrahedral substitutions, are rational in Z_1; that, similarly, all rational functions of z, which remain unaltered for the substitutions of the quadratic group, are rational in Z_2, &c. Hence (if we further consider the degree of the functions in question) Z is a rational function of Z_1 of the second degree, this again a rational function of Z_2 of the third degree, Z_2 in its turn a rational function of Z_3 of the second degree, and Z_3 itself, as is already noted in § 7, a rational function of z of the second degree. A glance at our earlier formulæ suffices to actually construct these rational functions. We find in order:

$$\frac{-12Z_1}{(Z_1-1)^2}=Z, \tag{8}$$

$$\left(\frac{Z_2-\alpha}{Z_2-\alpha^2}\right)^3=Z_1, \quad \left(\alpha=\frac{1+\sqrt{-3}}{2}\right), \tag{9}$$

$$-\frac{(Z_3-1)^2}{4Z_3}=Z_2, \tag{10}$$

and finally, as is self-evident:

$$\left(\frac{z_1}{z_2}\right)^2=Z_3. \tag{11}$$

It is just these formulæ, in which we now consider Z_1, Z_2, Z_3, *and* z *in order as the unknown, which are the auxiliary equations we sought.* It will be observed in particular, that the cubic auxiliary equation (9), as we had proposed, only needs a cube root for its solution.*

The *tetrahedral equation* is, without further trouble, solved by the way in these formulæ. In fact, we need only, in order to deal with it, allow the sequence of auxiliary equations to begin with (9). But the general *dihedral equation* also:

$$-\frac{(z_1^n-z_2^n)^2}{4z_1^nz_2^n}=Z, \tag{12}$$

offers no difficulties; in order to reduce it to a binomial equation, we need only, exactly as we did just now in the case of the quadratic group, introduce as the new unknown,

$$\left(\frac{z_1}{z_2}\right)^n=Z_1. \tag{13}$$

* The appearance of the irrationality α in (9), which distinguishes it from the rest, is the equivalent of the fact that, in order to reduce a cyclic equation of the third degree to the binomial form, we have always, as we remarked just now, to bring α to our assistance.

We have then for Z_1 the quadratic equation:

$$(14) \qquad -\frac{(Z_1-1)_2}{4Z_1}=Z,$$

and afterwards calculate $\frac{z_1}{z_2}$ from the binomial equation (13).

§ 8. The Resolvents of the Fifth Degree for the Icosahedral Equation.

Turning now to the icosahedral equation, we investigate first, and in detail, the resolvents of the fifth degree. We here use from the first the same fundamental theorems as were applied in the preceding paragraph. For the individual tetrahedral group contained in the icosahedral group a triply infinite number of rational functions of the twelfth degree of z remain unaltered, as we previously ascertained, which functions express themselves linearly in terms of any one of them, which we will call r, but which we will only fully define later on. Introducing this r as the unknown, the required resolvent of the fifth degree takes the form:

$$(15) \qquad F(r)=Z,$$

where F is a rational function of the fifth degree with numerical coefficients, and Z is the right side of the icosahedral equation. Our object will be to determine F. This is, of course, at once attained if we establish r explicitly as a function of $\frac{z_1}{z_2}$, and take into account, besides, the left side of the icosahedral equation. Howbeit, the matter is somewhat more complicated than in the case of the preceding paragraph, and hence I prefer to develop in the following paragraph a method by which we can determine the value of $F(r)$ without recurring at all to the formulæ in z.*

By the side of this first, which might be called the *function-theory* method, another presents itself—the *invariant* method. This is connected with the homogeneous substitutions of z_1, z_2 and the corresponding forms which remain unaltered; it is related, therefore, in the first place, to the problem of the

* I have repeatedly used this method in Bd. xii of the Math. Annalen, p 175, and Bd. xiv, p. 141, 416, &c. (1877–78), in order to establish equations defined in an analogous way.

icosahedron, and we shall only in a supplementary manner change the results obtained from it into resolvents of the icosahedral equations.

We have in § 1 of the preceding chapter collected, for each of the homogeneous groups of substitutions there described, the complete system of the corresponding entirely invariant forms. For the 120 substitutions of the homogeneous icosahedral group these are the forms f, H, T, themselves. On the other hand, for the 24 substitutions of the homogeneous tetrahedral group there are the corresponding *octahedral form*, the associated *cube* W, and a form of the twelfth degree, χ, for which, however, we can now put f, which is a linear combination of t^2 and χ. The most general entirely invariant tetrahedral form is therefore an arbitrary integral function of t, W, and f (homogeneous in z_1, z_2).

Let G be such a form. Assuming that it does not, at the same time, remain unaltered for the icosahedral substitutions, we obtain from it, by means of the icosahedral substitutions, 5 different forms, which we will denote by $G_0, G_1, \ldots G_4$. We construct the product:

$$\prod_{\nu} (G - G_\nu).$$

Here the coefficients of the different powers of G are symmetric functions of the G_ν's, *i.e.*, icosahedral forms. *Hence G will satisfy an equation of the fifth degree:*

$$(16) \qquad G^5 + aG^4 + bG^3 + cG^2 + dG + e = 0,$$

in which the coefficients a, b, . . . are integral functions of the f, H, T. The calculation of these coefficients is achieved immediately. For since we know the degree of the G_ν's in z_1, z_2, we know *a priori* that $a, b, c, \ldots$ can be composed linearly only of determinate combinations of f, H, T, finite in number, and all that is then required in order to determine the remaining unknown numerical coefficients is a comparison of a few terms in the explicit formulæ for f, H, T, and G.

In order now to transform the equation (16) into a resolvent of the icosahedral equation, we will multiply G, or divide, respectively, by such powers of f, H, T, that there results a rational function of null degree of z_1,z_2, *i.e.*, a rational function of z. We have then simply to introduce this function in (16) as the unknown, instead of G, whereupon the coefficients a, b, c, . . . will be of themselves transformed into rational functions of Z.

So much for the invariant-theory method.* To carry it out, I first compute in § 10 the explicit values of t and W. I then give in §§ 11, 12 the prepared equations on which t, on the one hand, on the other an arbitrary linear combination of W and $t\,W$, depends, equations which can then at once be changed into resolvents of the icosahedral equation. The first of these equations is also especially remarkable, because it has already occurred (certainly on quite different assumptions) in the earliest researches of *Brioschi* † on the solution of equations of the fifth degree, as we shall have to describe more in detail later on. The other will play an important rôle in our theory of the principal equation of the fifth degree, which we shall develop in Chapter II. of the following Part, and may therefore be here at once described as the *canonical resolvent*.‡ In § 13 I then explain, further, how these new resolvents of the fifth degree are connected with the resolvent of the r's (which was furnished by the function-theory method), and finally determine (in § 14) for it the value of the particular product of differences, which, as we know, must be rational in Z.

§ 9. The Resolvent of the r's.

To compute the resolvent of the r's (15), we first of all split up $F(r)$ into numerator and denominator, and take the particular value $Z=1$ into consideration, and therefore write, instead of (15):

$$(17) \qquad \phi(r) : \psi(r) : \chi(r) = Z : Z-1 : 1,$$

* I gave this, in the form here used, first in Bd. xii of the Math. Ann. (1877), p. 517, &c.

† See "Annali di Matematica," Ser. I, t. i, 1858.

‡ I first communicated the canonical resolvent, in a somewhat less simple form, however, in Bd. xii of the Annalen, p. 525. It is also implicitly the foundation of the parallel investigations of Gordan, which we shall describe in detail in the following Part (see in particular Bd. xiii of the Annalen, "Ueber die Auflösung der Gleichungen 5 Grades," 1878).

where ϕ, ψ, χ will be *integral* functions of the fifth degree. Now combining the original icosahedral equation:

$$H^3(z) : - T^2(z) : 1728 f^5(z) = Z : Z-1 : 1,$$

with this, we remark that $\phi=0$, $\psi=0$, $\chi=0$, give respectively those values of r which are inserted for the 20, 30, and 12 points of $H=0$, $T=0$, and $f=0$. The consideration of the figure gives us accordingly certain theorems concerning the linear factors of ϕ, ψ, χ.

It is clear, in the first place, that the aggregate of the points $f=0$ will be permuted amongst themselves for the 12 rotations which leave r unaltered (*i.e.*, for the 12 rotations of the corresponding tetrahedral group). Therefore r will assume the same value for all points of $f=0$. *Hence* $\chi(r)$ *is necessarily the fifth power of a linear expression.* We consider further the 30 points $T=0$. Amongst these are found, above all, the 6 summits of the octahedron belonging to the tetrahedral group (which we just now denoted by t). The remaining 24 points are divided (as is evident on a model) in virtue of the tetrahedral rotations into twice 12 associated ones. *We hence conclude that* $\psi(r)$ *contains one simply linear factor, and two others counted twice.* As regards these multiplicities, let us remark that $\psi(r)=0$, corresponding to the term $T^2(z)$ of the icosahedral equation, must represent the aggregate of points under consideration, counted twice. The linear factor, however, which vanishes at the 6 octahedral summits, will be of itself twice equal to zero; it need therefore be only counted once as contained in $\psi(r)$. On the other hand, the two other linear factors, on the same grounds, vanishing as they do in sets of 12 different points, and therefore only once vanishing, must occur in ψ counted twice. This agrees with the fact that the one linear factor presenting itself in $\chi(r)$ is to be taken quintuply. We consider, finally, the points $\phi(r)=0$ or $H=0$. Among them are found, as we know beforehand, the 8 summits of the cube W appertaining to the tetrahedral group. These distribute themselves in virtue of the tetrahedral group into twice 4 coordinated points, of which each remains fixed for 3 tetrahedral rotations. We have, in addition to these, 12 more points of $H=0$, which in respect to the 12 tetrahedral rotations form a single group. *Hence we conclude that* $\phi(r)$ *possesses only* 3 *different linear factors, of which the two which correspond to $W=0$ occur simply, while the third occurs as a cube.*

Summing up, we have reached a result which expresses itself in the replacing of formulæ (17) by the following:

$$
(18)\qquad \begin{aligned} Z : Z-1 : 1 &= c\ (r-\alpha)^3\,(r^2-\beta r+\gamma) \\ &: c'\ (r-\delta)\ (r^2-\epsilon r+\zeta)^2 \\ &: c''\,(r-\eta)^5, \end{aligned}
$$

understanding by $\alpha, \beta, \gamma, \ldots, c, c,' c,''$ constants which are still unknown.

The determination of these constants is a problem which is only determinate when we have previously defined r in an unambiguous manner. Let r be one of the triply infinite number of rational functions of the twelfth degree, which remain unaltered for the rotations of the tetrahedral group. We will now put, in particular, $r=\frac{t^2}{f}$, understanding by t (as above) the octahedral form appertaining to the tetrahedral group. Here t should be so chosen that, when arranged in powers of z_1, z_2, it begins with the term $+z_1^6$ and has altogether real coefficients.* Then the first result is that, in (18), $c''(r-\eta)^5$ is equal to C (since it is only to vanish for $r=\infty$), and therefore c is to be put $=c'$ while δ vanishes. We have further that C is to be taken $=-1728c$. For $\frac{t^2}{f}$, in consequence of our convention, reduces itself, for a very large value of $\frac{z_1}{z_2}$, to $\frac{z_1}{z_2}$ as a first approximation, while Z (in virtue of the icosahedral equation is to be replaced by $\frac{-z_1^5}{1728z_2^5}$. Finally, it follows that all the coefficients in (18) will be real. *We have therefore now so simplified formula* (18) *that we can write:*

$$
(19)\qquad \begin{aligned} Z : Z-1 : 1 &= (r-\alpha)^3\,(r^2-\beta r+\gamma) \\ &: r\,(r^2-\epsilon r+\zeta)^2 \\ &: -1728, \end{aligned}
$$

understanding by $\alpha, \beta, \gamma, \epsilon, \zeta$ real constants.

* Both these conditions can be satisfied, as a glance at the figure shows. For on the one hand, each of the 5 octahedra occurring in connection with the icosahedron contains a term with z_1^6, because none has a summit at $z_2=0$, and, on the other hand, amongst these octahedra is found one which has the meridian of real numbers for its circle of symmetry.

Now α, β, γ, ϵ, ζ must in any case, in correspondence with this formula, be so determined that the following relation is identically true:

$$(20) \qquad (r-\alpha)^3(r^2-\beta r+\gamma)+1728=r(r^2-\epsilon r+\zeta)^2.$$

On treating this identity by appropriate means, we recognise that, with its help, α, β, γ, ϵ, ζ *are fully determined.* Namely, we have first, on putting in (20) $r=0$:

$$\alpha^3\gamma=+1728.$$

Then, on differentiating (20) with respect to r, we find further:

$$(r-\alpha)^2\left(5r^2-(2\alpha+4\beta)\,r+(\alpha\beta+3\gamma)\right)$$
$$=(r^2-\epsilon r+\zeta)\,(5r^2-3\epsilon r+\zeta),$$

or, since $(r^2-\epsilon r+\zeta)$ and $(r-\alpha)^2$ are necessarily prime to one another:

$$5\epsilon=2\alpha+4\beta, \quad 10\alpha=3\epsilon,$$
$$5\zeta=\alpha\beta+3\gamma, \quad 5\alpha^2=\zeta,$$

therefore (by eliminating ϵ, ζ):

$$11\alpha=3\beta, \quad 64\alpha^2=9\gamma,$$

and by combination with the relation first found:

$$\alpha^5=3^5.$$

But now α is to be *real.* Thus we have $\alpha=3$, and hence $\beta=11$, $\gamma=64$, $\epsilon=10$, $\zeta=45$. *The resolvent of the r runs therefore simply thus:*

$$(21) \qquad \begin{aligned} Z:Z-1:1 &= (r-3)^3\,(r^2-11r+64)\\ &:r\,(r^2-10r+45)^2\\ &:-1728.\end{aligned}$$

§ 10. Computation of the Forms t and W.

We now give a supplementary computation of the forms t and W, whereby, on the one hand, we attain to an explicit exposition of the connection between the quantity r, used in the preceding paragraph, and the $\frac{z_1}{z_2}$ of the icosahedral equation, and, on the other hand, obtain the necessary foundation for the invariant-theory method for the construction of resolvents.

We remarked in § 12 of the first chapter, that to that icosahedral group which we have here to consider belong the rotations:

$$T,\ U,\ TU,$$

to which we then made to correspond, in § 7 of the second chapter, the substitutions:

$$z' = \frac{(\epsilon^4-\epsilon)\,z+(\epsilon^2-\epsilon^3)}{(\epsilon^2-\epsilon^3)\,z-(\epsilon^4-\epsilon)},$$
$$z' = -\frac{1}{z},$$
$$z' = \frac{(\epsilon^2-\epsilon^3)\,z+(\epsilon-\epsilon^4)}{(\epsilon-\epsilon^4)\ \ z+(\epsilon^2-\epsilon^3)}.$$

We compute, in a homogeneous form, for the pairs of points which remain fixed for these substitutions the following equations:

$$z_1^2-2(\epsilon^2+\epsilon^3)\,z_1z_2-z_2^2=0,$$
$$z_1^2+z_2^2=0,$$
$$z_1^2-2(\epsilon+\epsilon^4)\,z_1z_2-z_2^2=0.$$

But now the octahedron t will be constructed with just these 3 *pairs of points.* On further reflecting that the form t must contain the term $+z_1^6$, we have, accordingly, for the latter:

$$(22)\qquad \begin{aligned} t(z_1, z_2) = (z_1^2+z_2^2)\cdot\left(z_1^2-2(\epsilon+\epsilon^4)z_1z_2-z_2^2\right) \\ \cdot\left(z_1^2-2(\epsilon^2+\epsilon^3)z_1z_2-z_2^2\right) \\ = z_1^6+2z_1^5z_2-5z_1^4z_2^2-5z_1^2z_2^4-2z_1z_2^5+z_2^6. \end{aligned}$$

If we now wish to compute the corresponding W, this can be done, according to our earlier developments, on establishing the *Hessian* form of $t(z_1, z_2)$. We may further agree, as it is convenient for our later calculation to do, that $W(z_1, z_2)$ is to contain the term $-z_1^8$. We have thus:

$$(23)\qquad \begin{aligned} W(z_1, z_2) = -z_1^8+z_1^7z_2-7z_1^6z_2^2-7z_1^5z_2^3 \\ +7z_1^3z_2^5-7z_1^2z_2^6-z_1z_2^7-z_2^8, \end{aligned}$$

and we have thus already achieved the first object of the present paragraph.

We now subject t and W to the operations:

$$S^\nu : z_1' = \pm\epsilon^{3\nu} \cdot z_1,\ z_2' = \pm\epsilon^{2\nu} \cdot z_2.$$

Thus arise, respectively, those five values which always come under consideration simultaneously in the case of our equations of the fifth degree, and which we will call t_ν, W_ν. We find:

$$(24)\qquad \begin{aligned} t_\nu(z_1, z_2) = \epsilon^{3\nu} z_1^6 + 2\epsilon^{2\nu} z_1^5 z_2 - 5\epsilon^\nu z_1^4 z_2^2 \\ - 5\epsilon^{4\nu} z_1^2 z_2^4 - 2\epsilon^{3\nu} z_1 z_2^5 + \epsilon^{2\nu} z_2^6, \end{aligned}$$

$$(25)\qquad \begin{aligned} W_\nu(z_1, z_2) = -\epsilon^{4\nu} z_1^8 + \epsilon^{3\nu} z_1^7 z_2 - 7\,\epsilon^{2\nu} z_1^6 z_2^2 - 7\,\epsilon^\nu z_1^5 z_2^3 + 7\,\epsilon^{4\nu} z_1^3 z_2^5 \\ - 7\,\epsilon^{3\nu} z_1^2 z_2^6 - \epsilon^2 z_1 z_2^7 - \epsilon^\nu z_2^8. \end{aligned}$$

Here we will inquire expressly how the five t_ν's or W_ν's are permuted under the 120 homogeneous icosahedral substitutions. This, however, is derived already from the statement which we have made in § 8 of the first chapter concerning the corresponding geometrical figures; but it seems useful to connect the rule in question explicitly with our present formulæ. We have generated the 120 homogeneous icosahedral substitutions from the following formulæ by repetition and combination:

$$\begin{aligned} S: &\quad z_1' = \pm\epsilon^3 z_1,\ z_2' = \pm\epsilon^2 z_2, \\ T: &\ \pm\sqrt{5} \cdot z_1' = -(\epsilon - \epsilon^4)\, z_1 + (\epsilon^2 - \epsilon^3) z_2, \\ &\ \pm\sqrt{5} \cdot z_2' = +(\epsilon^2 - \epsilon^3)\, z_1 + (\epsilon - \epsilon^4) z_2. \end{aligned}$$

Introducing now these values of z_1', z_2', instead of z_1, z_2, in the forms t_ν (or the W_ν), new forms t'_ν arise, whose connection with the original t_ν's is given, after a little calculation, by:

$$(26)\qquad \begin{cases} S:\ t_\nu' = t_{\nu+1} \\ T:\ t_0' = t_0,\ t_1' = t_2',\ t_2' = t_1,\ t_3' = t_4,\ t_4' = t_3. \end{cases}$$

Here, in the formula for S, the indices are taken with respect to the modulus 5.

§ 11. The Resolvent of the u's.

We next compute the equation of the fifth degree which our t_ν's satisfy. If we write, in correspondence with the formula (16):

$$t^5 + at^4 + bt^3 + ct^2 + dt + e = 0;$$

a, b, c, . . . will be respectively of the 6th, 12th, 18th, . . . degrees. Now, they are at the same time to be integral functions of f, H, T. Hence a and c must in any case vanish, while b, d, e will be respectively proportional to f, f^2, T. Our equation of the fifth degree will therefore have the following form:

$$t^5 + \kappa f \,.\, t^3 + \lambda f^2 \,.\, t + \mu T = 0,$$

where κ, λ, μ are numerical factors. To determine them, we either introduce the value of t (22) and the values of f, H, T, as we exhibited them above, into this equation, arrange them in the order z_1^{30}, $z_1^{29}z_2$, . . . and require that the three highest terms which do not vanish identically shall be reduced to zero by appropriate values of κ, λ, μ; or, again, we determine, in the appropriate symmetric functions of the t_ν's (24), the highest term which does not vanish in each case, and compare this with the highest term in f, f^2, T. In both cases we get:

$$\kappa = -10,\ \lambda = 45,\ \mu = -1\,;$$

and our equation of the fifth degree thus runs as follows:*

(27) $$t^5 - 10f \,.\, t^3 + 45f^2 \,.\, t - T = 0.$$

In order now to pass to a resolvent of the icosahedral equation, we put, say,

(28) $$u = \frac{12f^2 \,.\, t}{T},$$

(where now u depends on $\frac{z_1}{z_2}$ alone). Thus we have, by a simple substitution:

(29) $$48u^5(1-Z)^2 - 40u^3(1-Z) + 15u - 4 = 0.$$

I shall henceforward denote this equation as the *resolvent of the u's.*

§ 12. The Canonical Resolvent of the Y's.

In our later researches on equations of the fifth degree, those equations in which the fourth and the third power of the unknown are wanting simultaneously will play a particularly important rôle. The equation of the fifth degree, which our W_ν's satisfy, evidently belongs to them. For we have $\Sigma W_\nu = 0$, $\Sigma W_\nu^2 = 0$, inasmuch as there are no icosahedral forms of

* This is just that equation which, as has been already explained, occurs in the early works of *Brioschi.*

degree 8 or 16. In just the same way the equation of the fifth degree belongs to them, which we can establish for the next highest tetrahedral form, $t \,.\, W$. For we have again identically (and on the same grounds): $\Sigma\, t_\nu W_\nu = 0$, $\Sigma\, (t_\nu W_\nu)^2 = 0$. But $\Sigma\, (W_\nu) \,.\, (t_\nu W_\nu)$ will also, in virtue of the same considerations, be identically zero. *Hence, generally, to our equations of the fifth degree those ones will belong whose roots are linear combinations of the W_ν's and the $t_\nu W_\nu$'s with constant coefficients:*

$$(30) \qquad Y_\nu = \sigma \,.\, W_\nu + \tau \,.\, t_\nu W_\nu.$$

We accordingly set ourselves the task of calculating out, for any values of σ, τ, the corresponding equation of the fifth degree. Inasmuch as the details of the calculation offer nothing of special interest, communicate the result immediately. We find:

$$(31) \qquad \begin{aligned} &Y^5 + 5Y^2(8f^2 \,.\, \sigma^3 + T \,.\, \sigma^2\tau + 72f^3 \,.\, \sigma\tau^2 + fT \,.\, \tau^3) \\ &\quad + 5Y(-fH \,.\, \sigma^4 + 18f^2H \,.\, \sigma^2\tau^2 + HT \,.\, \sigma\tau^3 + 27f^3H \,.\, \tau^4) \\ &\quad + (H^2 \,.\, \sigma^5 - 10fH^2 \,.\, \sigma^3\tau^2 + 45f^2H^2 \,.\, \sigma\tau^4 + TH^2 \,.\, \tau^5) = 0. \end{aligned}$$

In order to construct herefrom a resolvent of the icosahedral equation, we have but to recur to the formula (28), and put instead:

$$(32) \qquad v = \frac{12f \,.\, W}{H}.$$

Then we can write formula (30) as follows:

$$(33) \qquad Y_\nu = m \,.\, v_\nu + n \,.\, u_\nu v_\nu,$$

where m is put $= \dfrac{\sigma \,.\, H}{12f}$, $n = \dfrac{\tau \,.\, HT}{144f^3}$.

On introducing into (31) the values of σ, τ, resulting herefrom, we obtain:

$$(34) \qquad \begin{aligned} &Z \,.\, Y^5 + 5Y^2\left(8m^3 + 12m^2n + \frac{6mn^2 + n^3}{(1-Z)}\right) \\ &\quad + 15Y\left(-4m^4 + \frac{6m^2n^2 + 4mn^3}{(1-Z)} + \frac{3n^4}{4(1-Z)^2}\right) \\ &\quad + 3\left(48m^5 - \frac{40m^3n^2}{(1-Z)} + \frac{15mn^4 + 4n^5}{(1-Z)^2}\right) = 0. \end{aligned}$$

This is that resolvent of the fifth degree of the icosahedral equation which we shall later on refer to as the *canonical resolvent.*

§ 13. Connection of the New Resolvent with the Resolvent of the r's.

We have now to exhibit the connection of our new resolvent with the resolvent of the r's (§ 10).

First, as regards the agreement of the function-theory and invariant-theory methods, we write the equation (27), say, as follows:

(35) $$T = t(t^4 - 10ft^2 + 45f^2);$$

now squaring, dividing both sides by f^5, and finally writing r again for $\frac{t^2}{f}$, we have:

$$-1728\,(Z-1) = r(r^2 - 10r + 45)^2,$$

an equation which, in fact, is identical with (21).

We shall have further to express

$$u = \frac{12t\,.\,f^2}{T} \text{ and } v = \frac{12\,W\,.\,f}{H}$$

rationally in terms of r.

As regards u, we effect this at once on introducing for T the value (35).

We thus find:

(36) $$u = \frac{12}{r^2 - 10r + 45}.$$

To exhibit v similarly, let us recall that, according to the developments of § 10, the points $\frac{H}{W} = 0$ are at the same time represented by $r - 3 = 0$. Therefore $\frac{H}{W}$ will be identical with $t^2 - 3f$, save as to a factor. The comparison of any term in the development in terms of z_1, z_2, shows that this factor $= +1$. Hence we have without further proof:

(37) $$v = \frac{12}{r-3}.$$

Finally, introducing the values of (36), (37), in (33), we have:

(38) $$Y_\nu = \frac{12m\,(r-3) + 144n}{(r-3)\,(r^2 - 10r + 45)}.$$

We should, of course, be now able to compute also the resolvent of the u's and the canonical resolvent, on eliminating r between (21) and (36) or (38).*

§ 14. On the Products of Differences for the u's and the Y's.

We now further calculate, also in view of their later applications, the products of differences of the u's and the Y's which, as we know, are rational in Z. We consider, say, first the following product:

$$\prod_{\nu<\nu'} (t_\nu - t_{\nu'}),$$

where the symbol under the product-sign is to denote that only those 10 factors are to be multiplied out for which ν is $<\nu'$ (while ν and ν' are to be simultaneously susceptible of the values 0, 1, 2, 3, 4). This product is known to be equal to the determinant:

$$\begin{vmatrix} 1 & t_0 & . & . & . & t_0^4 \\ 1 & t_1 & . & . & . & t_1^4 \\ . & . & . & . & . & . \\ . & . & . & . & . & . \\ 1 & t_4 & . & . & . & t_4^4 \end{vmatrix}$$

We now multiply this last expression, according to the rules of multiplication of determinants, by:

$$\begin{vmatrix} 1 & 1 & 1 & 1 & 1 \\ 1 & \epsilon & \epsilon^2 & \epsilon^3 & \epsilon^4 \\ 1 & \epsilon^2 & . & . & . \\ 1 & \epsilon^3 & . & . & . \\ 1 & \epsilon^4 & . & . & . \end{vmatrix} = 25\sqrt{5}, \left(\epsilon = e^{\frac{2i\pi}{5}}\right).$$

Thus arises a new determinant with real and integral numerical coefficients throughout:

* This is the manner in which *Herr Kiepert* has derived the canonical resolvent: "Auflösung der Gleichungen fünften Grades" (Göttinger Nachrichten of the 6th of July 1878; Borchardt's Journal, Bd. 79).

$$5 \,.\, \begin{vmatrix} \sum \epsilon^{\nu} t_{\nu} & \sum \epsilon^{\nu} t^2_{\nu} & \sum \epsilon^{\nu}{}_{\nu} t^3_{\nu} & \sum \epsilon_{\nu} t^4_{\nu} \\ \sum \epsilon^{2\nu} t_{\nu} & \sum \epsilon^{2\nu} t^2_{\nu} & \cdot & \cdot \\ \sum \epsilon^{3\nu} t_{\nu} & \cdot & \cdot & \cdot \\ \sum \epsilon^{4\nu} t_{\nu} & \cdot & \cdot & \cdot \end{vmatrix}$$

This is of the 60th degree in z_1, z_2; it will therefore (as an icosahedral form) be necessarily equal to a linear combination of H^3 and f^5. *On actually calculating the terms which contain* z_1^{60} *and* $z_1^{55} z_2^{5}$, *we verify that this*

$$= 5^5 \,.\, H^3\,(z_1, z_2).$$

Hence our original product of differences:

$$\prod_{\nu < \nu'} (t_{\nu} - t_{\nu'}) = 25\sqrt{5} \,.\, H^3\,(z_1, z_2).$$

But now we have:

$$t_{\nu} = \frac{T \,.\, u_{\nu}}{12 f^2}.$$

Hence the product of the differences of the u's will be:

$$\prod_{\nu < \nu'} (u_{\nu} - u_{\nu'}) = -\frac{25\sqrt{5}}{144} \cdot \frac{Z}{(Z-1)^5}. \tag{39}$$

In a similar manner I compute the product of differences of the Y's. Starting first from (30), we find:

$$\begin{aligned} \prod_{\nu < \nu'} (Y_{\nu} - Y_{\nu'}) = -25\sqrt{5} \,.\, H\{ & T^2 \,.\, \sigma^{10} + 2^4 \,.\, 5 \,.\, 7 \,.\, f^3 T \,.\, \sigma^9 \tau \\ & + 5^2 \,.\, f(2^6 \,.\, 3^4 \,.\, f^5 - H^3)\sigma^8\tau^2 + 2^6 \,.\, 3^3 \,.\, 5 \,.\, f^4 T \,.\, \sigma^7 \tau^3 \\ + 2 \,.\, 5 \,.\, f^2(2^6 \,.\, 3^4 \,.\, 7 \,.\, f^5 - 31 \,.\, H^3)\sigma^6\tau^4 & - T(2^5 \,.\, 3^4 \,.\, 7 \,.\, f^5 + 11 \,.\, H^3)\sigma^5\tau^5 \\ - 2 \,.\, 3^2 \,.\, 5 \,.\, f^3(2^6 \,.\, 3^4 \,.\, 7 \,.\, f^5 - 13 \,.\, H^3)\sigma^4\tau^6 & - 2 \,.\, 5^2 \,.\, fT(2^5 \,.\, 3^4 \,.\, f^5 - H^3)\sigma^3\tau^7 \\ - 3^4 \,.\, 5 \,.\, f^4(2^6 \,.\, 3^3 \,.\, 7 \,.\, f^5 - 11 \,.\, H^3)\sigma^2\tau^8 & - 3^2 \,.\, 5 \,.\, f^2 T(2^4 \,.\, 3^4 \,.\, f^5 - H^3)\sigma\tau^9 \\ & - (2^6 \,.\, 3^6 \,.\, 11 \,.\, f^{10} - 3^4 \,.\, 7 \,.\, f^5 H^3 + H^6)\tau^{10}\}, \end{aligned}$$

and then, passing to (33), we have the final result:

$$\prod_{\nu<\nu'}(Y_\nu - Y_{\nu'}) = \frac{-25\sqrt{5}}{Z^3}\Big\{\, 2^8 \,.\, 3^4\,(1-Z)\,m^{10} + 2^8 \,.\, 3^2 \,.\, 5 \,.\, 7 \,.\, m^9 n$$

$$+\frac{2^6 \,.\, 3^3 \,.\, 5^2\,(3-Z)\,m^8n^2 + 2^8 \,.\, 3^4 \,.\, 5 \,.\, m^7n^3}{1-Z}$$

$$+\frac{2^5 \,.\, 3^2 \,.\, 5\,(3 \,.\, 7 - 31 \,.\, Z)\,m^6n^4 - 2^5 \,.\, 3^3\,(3 \,.\, 7 + 2 \,.\, 11 \,.\, Z)\,m^5n^5}{(1-Z)^2}$$

(40)

$$-\frac{2^3 \,.\, 3^3 \,.\, 5\,(3 \,.\, 7 - 13 \,.\, Z)\,m^4n^6 + 2^4 \,.\, 3^2 \,.\, 5\,(3-2Z)\,m^3n^7}{(1-Z)^3}$$

$$-\frac{3^4 \,.\, 5\,(7 - 11 \,.\, Z)\,m^2n^8 + 3^3 \,.\, 5\,(3 - 2^2 \,.\, Z)\,m\,n^9}{(1-Z)^4}$$

$$-\frac{(3^2 \,.\, 11 - 3^3 \,.\, 7 \,.\, Z + 2^6 \,.\, 3^2 \,.\, Z^2)\,n^{10}}{2^2 \,.\, (1-Z)^6}\Big\}.$$

§ 15. The Simplest Resolvent of the Sixth Degree.

At the conclusion of this chapter, and with the special object of forming later on a simple connection between our own developments and the earlier investigations of other mathematicians, let us further consider the simplest resolvent of the sixth degree of the icosahedral equation, and, in fact, with this object, we will at once employ the invariant-theory method.*

Among the six dihedral groups containing each 10 rotations, which here come under consideration, let us select that one whose principal axis joins the two points $z=0$, ∞. The lowest form, which remains completely unaltered for the corresponding homogeneous substitutions, is, as we know beforehand, the *square* of z_1z_2. We shall therefore first compute the equation of the sixth degree which this square, or rather the quantity:

$$(41) \qquad \phi_\infty = 5z_1^2z_2^2$$

satisfies, where the numerical factor 5 has been advisedly attached, and the index ∞ affixed to the ϕ, in order to have the symbols $\phi_\nu(\nu=0, 1, 2, 3, 4$ [mod. 5]) available for the corresponding expressions which appertain to the remaining 5 icosahedral diagonals. On applying to ϕ_∞ the homogeneous icosahedral substitutions which correspond to TS^ν, we find for these ϕ_ν's:

$$(42) \qquad \phi_\nu = (\epsilon^\nu z_1^2 + 2z_1z_2 - \epsilon^{4\nu}z_2^2)^2.$$

* Compare Math. Annalen again, Bd. xii, pp. 517, 518.

Now let the equation of the sixth degree, which the ϕ's satisfy, be:

$$\phi^6+a'\phi^5+b'\phi^4+c'\phi^3+d'\phi^2+e'\phi+f'=0,$$

then a', b', c', . . . are icosahedral forms of the 4th, 8th, 12th, . . . degrees respectively. Hence it follows at once that $a'=b'=d'=0$, while c', e', f', must, apart from numerical factors, coincide with f, H, and f^2 respectively. We determine these factors in the well-known manner by returning to the values of f, H, and ϕ in z_1, z_2. *We thus find with little trouble the following equation:*

(43) $$\phi^6-10f\,.\,\phi^3+H\,.\,\phi+5f^2=0.$$

Let us now pay attention for a moment to the *group* of this equation. This will be given, as follows from our earlier developments, by those 60 permutations of the ϕ's which correspond to the 120 homogeneous icosahedral substitutions (where we must not forget that we have adjoined ϵ once for all). Now the latter are all composed of the substitutions S and T, which we again exhibited in § 10. Clearly the ϕ_∞ remains unaltered for S, while ϕ_ν is transformed into $\phi_{\nu+1}$. We can compress this into the single formula:

$$\nu'\equiv\nu+1 \pmod{5},$$

inasmuch as for $\nu=\infty$ the ν' so determined will also be ∞. On the other hand, for T, ϕ_∞ will be interchanged with ϕ_0, ϕ_1 with ϕ_4, ϕ_2 with ϕ_3, which will be repeated by means of the single formula:

$$\nu'\equiv-\frac{1}{\nu} \pmod{5}.$$

In the two formulæ so obtained, *all* formulæ:

$$\nu'\equiv-\frac{a\nu+\beta}{\gamma\nu+\delta} \pmod{5},$$

are now comprehended, according to known theorems of the theory of numbers, where a, β, γ, δ, are integral numbers which satisfy the congruence $(a\delta-\beta\gamma)\equiv1 \pmod{5}$. In fact, the number of these formulæ is equal to 60, so long as we always count as only one all such systems of values a, β, γ, δ, as are identical for the modulus 5, or are brought to identity by a uniform change of sign. Therefore:

The group of our equation of the sixth degree will be constructed with those 60 *permutations of the roots* ϕ_ν, *which are provided by the various formulæ:*

$$(44) \qquad \nu' \equiv \frac{\alpha\nu + \beta}{\gamma\nu + \delta} \text{ (mod. 5)}.$$

But this is, according to the researches of Galois, exactly that group which belongs to the modular equation of the sixth order, for a transformation of the fifth degree of elliptic functions. And, in fact, Herr Kronecker, who started from elliptic functions, while following up incidental statements of Jacobi's, had long before deduced exactly the same equation (43), though, of course, with a different notation.* We shall return again, and in some detail, to this circumstance.

In order now to transform (43) into a resolvent of the icosahedral equation, let us put, say,

$$(45) \qquad \zeta = \frac{\phi \,.\, H}{12f^2};$$

we thus obtain by mere substitution:

$$(46) \qquad \zeta^6 - 10Z \,.\, \zeta^3 + 12Z^2 \,.\, \zeta + 5Z^2 = 0.$$

We will amplify this result by deriving from it a second resolvent, whose root is a rational function of the 10th degree of z, which does not alter for the 10 substitutions of the dihedral group under consideration. Such a function is, for example:

$$(47) \qquad \xi = \frac{\phi_\nu^3}{f},$$

since, namely, the numerator and denominator of this expression have in common the factor $\sqrt{\phi}$, quadratic in z_1, z_2. In order to construct the corresponding resolvent, let us write (43) in the following manner:

$$(48) \qquad -H = \frac{\phi^6 - 10f \,.\, \phi^3 + 5f^2}{\phi},$$

then cube, and divide both sides by $1728f^5$. We have thus:

$$Z = \frac{(\xi^2 - 10\xi + 5)^3}{-1728\xi},$$

* See the passages cited in the first and third chapters of the following Part, and compare, on the other hand, § 8 of the chapter following this.

or, also (if we rearrange the value of $(Z-1)$ properly):

$$(49)\qquad \begin{aligned} Z : Z-1 : 1 = &(\xi^2-10\xi+5)^2 \\ &: (\xi^2-4\xi-1)^2\,(\xi^2-22\xi+125) \\ &: -1728\xi. \end{aligned}$$

We should also have been able to derive this equation from function-theory considerations, without any use of explicit formulæ.*

I further give, in conclusion, the formulæ by means of which $\zeta(45)$ is expressed rationally in terms of our present ξ. According to (48):

$$\zeta = \frac{\phi H}{12f^2} = \frac{-\phi^6+10f\,.\,\phi^3-5f^2}{12f^2},$$

and therefore:

$$(50)\qquad \zeta = \frac{-\xi^2+10\xi-5}{12}.$$

§ 16. Concluding Remarks.

The developments of the last paragraphs have manifold relations with the applications which are to be made of them in the part here following.

I may point out at once that the considerations of the present chapter will be of the weightiest importance for our further process of thought. Let me state this more precisely.

We have already seen, in the third chapter of the present Part, that we can consider the solution of our fundamental equation, from a function-theory point of view, as a generalisation of the elementary problem; *to extract the nth root from a magnitude Z.* The algebraical reflexions of the present chapter have then shown us that the irrationalities which are introduced by the equations of the dihedron, tetrahedron, and octahedron can be computed by repeated extractions of roots. *The icosahedral irrationality, on the contrary, has maintained its independent significance.* Hence an extension of the ordinary theory of equations seems to be indicated. In the latter we are generally restricted to the investigation of those problems which

* Compare Mathematische Annalen, xiv, p. 143 (formula (19) of that page).

admit of solution by repeated extraction of roots. *We will now adjoin, as a further possible operation, the solution of the icosahedral equation, and ask whether, among the problems which do not admit of solution by mere extraction of roots, there may not be some for which this can be effected by the help of the icosahedral irrationality.*

In this sense our second Part now deals with the general problem of the solution of equations of the fifth degree. The attempt to accomplish this solution with the help of the icosahedral equation appears the more natural, inasmuch as the group of the equations of the fifth degree, after adjunction of the square root of the discriminant, is simply isomorphic with the group of the icosahedral equation, and as we have, in the resolvents of the fifth degree of the icosahedral equation (previously established), just so many special equations of the fifth degree whose relation to the icosahedral equation is *a priori* established.

CHAPTER V

GENERAL THEOREMS AND SURVEY OF THE SUBJECT

§ 1. Estimation of our Process of Thought so far, and Generalisations thereof.

Having now, in the third and fourth chapters, studied the essential properties of our fundamental problem, we will inquire where lies the proximate cause of the remarkable simplicity which has manifested itself therein all along. About this, I believe, there can be no doubt, viz., *it is the property of our problems that from one of their solutions the others always proceed by means of linear substitutions which are* a priori *known.* The geometrical apparatus, from which we started in the developments of the first and second chapters, has served to lead up to our problems, and to illustrate their primary properties; now it has done us this service, we can henceforward leave it on one side.* Having grasped this notion, we shall naturally ask if there may not exist other equations, or systems of equations, also, which agree in that most essential point with our fundamental problem.

We therefore first seek, so far as it is possible, for new finite groups of linear substitutions of a variable z (or two homogeneous variables z_1, z_2). But we will show immediately (§ 2) that all such groups return to the ones already known to us. If we, therefore, conceive our statement of the question in the

* This is only meant to apply *ad hoc*, and for the developments of the second Part here following. For carrying out more thoroughly the generalisations proposed in the text, an illustrative notation, at all events when we have to deal with transcendental functions, is for the time quite indispensable, as also in § 6 of the present chapter, where we involuntarily, so to say, return to geometrical explanations.

obvious manner explained, the equations and systems of equations hitherto treated of are the only ones of their kind. This is a result which is calculated to attach a certain absolute value to our previous considerations, which, on account of their inductive form, at first appear to aim at no definite object. In fact, we see that our fundamental equations occur as a specially remarkable circumscribed group among numerous mathematical investigations of the last few years. In regard to this, I shall effect in § 3 following, the simple developments by means of which we show that, with the help of our fundamental equations, *all linear homogeneous differential equations of the second order with rational coefficients, which have entirely algebraical integrals*, can be established with little trouble. I refer, however, for the analogous significance of our fundamental equations for the linear homogeneous differential equations of the nth order with rational coefficients, to the memoir of *Halphen* * already quoted; further, as concerns the rôle which our fundamental equations play in the theory of elliptic modular functions, and similarly in the investigation, by the theory of numbers, of binary quadratic forms, to my own investigations † and those of *Herr Gierster*.‡ Meanwhile we can generalise our statement of the question in a twofold sense.

In the first place, we can, instead of the variables z_1, z_2, take into consideration *a larger number of homogeneous variables*, $z_1, z_2, \ldots z_n$, and inquire for the finite groups of linear substitutions which may exist in their case. I will presently (in §§ 4, 5) treat this more fully, and will here only observe that, as a consequence of the views thus unfolding themselves, the developments of the second Part here following appear as a single contribution to a general theory, which embraces the whole theory of equations.

Our second generalisation proceeds in another direction: *we will retain the one variable* $z=\frac{z_1}{z_2}$, *but, on the other hand, take into consideration infinite groups of linear substitutions.* Here

* "Sur la reduction des équations différentielles linéaires aux formes intégrables." Mémoires présentés, &c., xxviii, 1 (1880–83).

† *Cf.* especially Bd. xiv of the Math. Ann., pp. 148–160 (1878).

‡ "Ueber Relationen zwischen Classenzahlen binärer quadratischer Formen von negativer Determinante," Erste Note (Göttinger Nachrichten of June 4, 1879, or Math. Ann., Bd. xvii, p. 71, &c.).

that vast region opens out, *single-valued transcendent functions, with linear transformations into themselves,* to which general attention has recently been drawn from various quarters, but particularly by *M. Poincaré.** It is, of course, impossible for me to enter more minutely into the questions connected with this matter in the following paragraphs. My exposition is only to carry us so far that the position of the simplest class of functions among the others, viz., *the elliptic modular functions,* may be clearly conceived. To this is attached the proof (§§ 7, 8) that the equations of the tetrahedron, octahedron, and icosahedron admit of solution by elliptic modular functions in a similar manner to that in which, say, a binomial equation is solved by logarithms, a cubic equation (and also the general equation of the dihedron) by trigonometric functions; and this proof I wished to bring forward in its general outline, because it marks that point on which in the theory of equations, and particularly of equations of the fifth degree, the interest of mathematicians has been continuously concentrated. We can, evidently, also combine the generalisations here suggested; we can study transcendental functions of *several* variables with *an infinite number* of linear transformations into themselves.† But more important for us here are, I think, the considerations which I develop in § 9, in consequence of which absolutely no material difference exists between the two kinds of generalisation. Hence the perspective to which the consideration in § 5 of the finite groups has already led us will be, so to say, extended to an infinite distance.

§ 2. Determination of all Finite Groups of Linear Substitutions of a Variable.

The problem of determining all possible finite groups of linear substitutions of a variable has been dealt with in various ways.

* *Cf.* the numerous communications of *Poincaré* in the "Comptes Rendus de l'Académie des Sciences," as well as his memoirs in Bd. xix of the Math. Annalen, and in Bd. i and ii of Acta Mathematica (1881–83). Moreover, my essay in Bd. xxi of the Math. Ann. (1882) may also be consulted: "Neue Beiträge zur Riemann'schen Functionentheorie:" there, particularly, the literature of the subject is noted and described in detail.

† The latest researches of M. Picard move in this direction; *cf.* Comptes Rendus, 1882–83, also Acta Mathematica, Bd. i and ii.

With my primary geometrical method * is connected the analytical method of *Herr Gordan*,† then the general treatment by *M. C. Jordan*,‡ by means of which he is in a position to solve the corresponding question for the case of a larger number of variables. I shall here use a method of consideration, based on the function-theory, which I have already incidentally pointed out. § This starts from the idea of taking into consideration at once the *equations*, whose roots will be transformed into one another by the substitutions of the group, whence it may easily be shown that these equations practically return to the fundamental equations hitherto investigated. The process of thought, on which *M. Halphen* has been lately engaged ‖ for a similar purpose, is not essentially different from the one here given. Moreover, a determination of all finite groups of linear substitutions of a variable is also implicitly contained in the investigations of *Herr Fuchs* on algebraically integrable differential equations of the second order,¶ investigations which we have already more than once cited in Chapters II and III, and to which we shall again pay regard in the following paragraph. We may say that these works of Herr Fuchs differ from mine in the fact that he brings forward the standpoint of the theory of forms quite at the beginning, while I commence with function-theory considerations.

Let

$$\psi_0(x)=x,\ \psi_1(x),\ \psi_2(x)\ \ldots\ldots\ \psi_{N-1}(x)$$

be the N linear functions, which, equated to x', represent a finite group of N linear substitutions of the variable x. Further, let a, b, be any two quantities, so chosen that none of the expressions $\psi(a)$ are equal to b, or, what is the same thing, none

* "Sitzungsberichte der Erlanger physikalisch-medicinischen Gesellschaft of July 1874," Math. Annalen, Bd. ix (1875).

† "Ueber endliche Gruppen linearer Substitutionen einer Veränderlichen," Math. Annalen, Bd. xii (1877).

‡ "Mém. sur les équations diff. lin. à intégrale algébrique," Borchardt's Journal, Bd. 84 (1878); also "Sur la déterm. des groupes d'ordre fini contenus dans le groupe linéaire," Atti della Reale Accad. di Napoli (1880).

§ Math. Annalen, Bd. xiv, pp. 149–150 (1878).

‖ *Vide* p. 125 (footnote).

¶ Göttinger Nachrichten of August 1875; Borchardt's Journal, Bd. 81, 85 (1875–77).

of the expressions ψ (b) are equal to a. We then form the equation:

$$(1)\qquad \frac{(\psi_0(x)-a)\,(\psi_1(x)-a)\,\ldots\ldots\,(\psi_{N-1}(x)-a)}{(\psi_0(x)-b)\,(\psi_1(x)-b)\,\ldots\ldots\,(\psi_{N-1}(x)-b)}=X.$$

Then we have evidently an equation of the Nth degree, which remains unaltered for the N substitutions of our group, and whose N roots, corresponding to an arbitrary value of X, therefore, in every case, proceed from one of themselves by our N substitutions. In fact, if we substitute in (1), instead of x, any ψ (x), the consequence is simply, since the ψ's by hypothesis form a group, that the factors in the numerator, and likewise the factors in the denominator of the left side of (1), are permuted with one another in a certain manner.

Our assertion will now be this: *that we shall be able to transform the equation* (1) *into one of the fundamental equations hitherto considered by us by simply substituting for x and X appropriate linear functions of x and X:*

$$z=\frac{\alpha x+\beta}{\gamma x+\delta},\quad Z=\frac{aX+b}{cX+d}.$$

To prove this, we first ask for what values of X the equation (1) may possess multiple roots. It is certain that if, for one value of X, one set of ν x-roots become identical, *then all the corresponding x-roots coincide in sets of ν.* This follows from the consideration of the substitutions ψ, just as we have proved the same theorem in the first chapter with respect to the groups of rotations, and those points on the sphere which remain fixed for certain rotations. We will now assume that to the values $X=X_1, X_2, \ldots$ only ν_1-tuple, ν_2-tuple $\ldots$ roots correspond in the sense explained. According to the explanations of § 4 of our third chapter, we have then for the functional determinant of the $(2N-2)^{\text{th}}$ degree, which is computed from the numerator and denominator of the left side of (1) [after we have turned both into integral functions of x by multiplication by the denominators of the ψ's], $\frac{N}{\nu_1}$ roots of multiplicity (ν_1-1), $\frac{N}{\nu_2}$ roots of multiplicity (ν_2-1), &c. Hence:

$$\sum\frac{N}{\nu_i}(\nu_i-1)=2N-2,$$

or otherwise written:

(2) $$\sum\left(1-\frac{1}{\nu_i}\right)=2-\frac{2}{N}.$$

Our method will now first consist in *considering this equation as a diophantine equation for the integral numbers* ν_i, N, *and seeking all the systems of solution thereof.*

This latter is done in an extremely simple manner. We first agree that the number of the ν_i's cannot be less than 2, nor greater than 3 (assuming, of course, N to be >1). Namely, if the number of the ν_i's were equal to 1, the left side of (3) would be <1, while the right side for $N>1$ is greater than or equal to 1. But if the number of the ν_i's were $\geqq 4$, the left side of (2) would be $\geqq 2$, because each element $\left(1-\frac{1}{\nu_i}\right)$ of the sum is itself $\geqq\frac{1}{2}$, and this would be no less a contradiction.

We now first take the number of the ν_i's equal to 2, and therefore simply write instead of (2)

$$\left(1-\frac{1}{\nu_1}\right)+\left(1-\frac{1}{\nu_2}\right)=\left(2-\frac{2}{N}\right),$$

or

$$\frac{1}{\nu_1}+\frac{1}{\nu_2}=\frac{2}{N}.$$

Now it is self-evident that none of the ν_i's can be $>N$; therefore $\frac{1}{\nu_i}\geqq\frac{1}{N}$. We hence conclude that in the above case $\frac{1}{\nu_1}$ and $\frac{1}{\nu_2}$ must both be equal to $\frac{1}{N}$. *Hence we have:*

(3) $$\nu_1=\nu_2=N,$$

where N is arbitrary; and this is our first system of solution.

Let us further take the number of the ν_i's equal to 3, and therefore put, instead of (2), the equation:

(4) $$\frac{1}{\nu_1}+\frac{1}{\nu_2}+\frac{1}{\nu_3}=1+\frac{2}{N}.$$

Then I say, in the first place: *at least one of the ν_i's must be equal to* 2. Namely, if each of the three were $\nu_i\geqq 3$, the left-hand side of (4) would be $\leqq 1$, which is impossible. We therefore put, say, $\nu_1 = 2$. For the remainder:

$$\frac{1}{\nu_2} + \frac{1}{\nu_3} = \frac{1}{2} + \frac{2}{N}.$$

It is now possible that a second ν, say ν_2, is equal to 2. We then find:

$$\frac{1}{\nu_3} = \frac{2}{N}.$$

Thus we have our second system of solution, which we will denote as follows, understanding by n an arbitrary number:

(5) $$N = 2n, \quad \nu_1 = 2, \quad \nu_2 = 2, \quad \nu_3 = n.$$

But if neither of the two numbers ν_2, ν_3, is equal to 2, at least one of them must be equal to 3. For otherwise $\frac{1}{\nu_2} + \frac{1}{\nu_3}$ would be $\leqq \frac{1}{2}$, whereas it is to be $> \frac{1}{2}$. Accordingly, let us put $\nu_2 = 3$. There remains:

$$\frac{1}{\nu_3} = \frac{1}{6} + \frac{2}{N}.$$

Therefore anyhow $\nu_3 < 6$. On the other hand, we can choose $\nu_3 = 3$, 4, 5, according as we wish. We get correspondingly $N = 12$, 24, 60, and then in each case our conditions are all satisfied. *There are therefore three more systems of solution, which are embraced in the following table:*

(6) $$\begin{cases} N = 12, \ \nu_1 = 2, \ \nu_2 = 3, \ \nu_3 = 3; \\ N = 24, \ \nu_1 = 2, \ \nu_2 = 3, \ \nu_3 = 4; \\ N = 60, \ \nu_1 = 2, \ \nu_2 = 3, \ \nu_3 = 5. \end{cases}$$

The five different systems of solution so found correspond exactly, as we see at once, to our five fundamental equations: the binomial equation, the equations of the dihedron, tetrahedron, octahedron, and icosahedron. *We will now show that, according to the system of solution* (3), *or* (5), *or* (6), *which we like to attribute to our diophantine equation, we can in fact in each case transform our equation* (1), *in the way proposed, into the corresponding fundamental equation.*

Let us take the case (3) to start with. Instead of X we may introduce in it:

$$Z = \frac{X - X_1}{X - X_2}.$$

We have then for $Z = 0$ and for $Z = \infty$ an N-fold root x. Our equation (1) therefore admits of being written as follows:

$$\left(\frac{x - x_1}{x - x_2}\right)^N = Z;$$

and here we have only to put:

$$\frac{x - x_1}{x - x_2} = z,$$

in order to have before us the *binomial equation:*

$$z^N = Z.$$

In the other cases we will choose:

$$Z = \frac{X - X_2}{X - X_3} \cdot \frac{X_1 - X_3}{X_1 - X_2},$$

so that for $Z = 0$ merely ν_2-tuple, for $Z = \infty$ merely ν_3-tuple, for $Z = 1$ merely ν_1-tuple, *i.e.*, double roots enter. Denoting by Φ_1, Φ_2, Φ_3 approximate integral functions of x, our equation (1) takes then the following form:

$$Z : Z - 1 : 1 = \Phi_2^{\nu_2}(x) : \Phi_1^{\nu_1}(x) : \Phi_3^{\nu_3}(x),$$

where we must suppose for ν_2, ν_1, ν_3, one of our systems of solution introduced. We now combine with this the corresponding fundamental equation to which we had previously given the form:

$$Z : Z - 1 : 1 = cF_2^{\nu_2}(z) : c'F_1^{\nu_1}(z) : F_3^{\nu_3}(z).$$

Our assertion will be proved if we show that, *in consequence of these two equations, z is a linear function of x:*

$$z = \frac{\alpha x + \beta}{\gamma x + \delta}.$$

To this end we recall the differential equation of the third order, which we previously established for z as a function of Z (see § 8 of Chapter III):

$$[z]_Z = \frac{\nu_1^2 - 1}{2\nu_1^2 (Z - 1)^2} + \frac{\nu_2^2 - 1}{2\nu_2^2 . Z^2} + \frac{\frac{1}{\nu_1^2} + \frac{1}{\nu_2^2} - \frac{1}{\nu_3^3} - 1}{2(Z - 1)Z}.$$

On going through the steps of the proof which we used in the establishment of this equation, we recognise that x *in each case satisfies the same differential equation with respect to* Z. Now all solutions of such a differential equation are, as we know, linear functions of any particular solution. Hence z is also a linear function of x, *q.e.d.*

We will sum up the result thus obtained more concisely. Our object is to seek for all finite groups of linear substitutions:

$$x' = \psi_i(x), \quad i = 0, 1, \ldots . (N-1),$$

We now recognise that we obtain all of them by choosing as our starting-point the finite groups which we collected in § 7 of the second chapter, and then introducing in the formulæ there given, instead of z, an arbitrary x, by means of the equation: $z = \frac{\alpha x + \beta}{\gamma x + \delta}$, whereupon, of course, z' will have to be replaced in a corresponding manner by $x' = \frac{-\delta z' + \beta}{\gamma z' - \alpha}$.

§ 3. Algebraically Integrable Linear Homogeneous Differential Equations of the Second Order.

Interrupting our general course of ideas, let us now concern ourselves, as we proposed in § 1, with the problem: *to present all linear homogeneous differential equations of the second order with rational coefficients:*

$$(7) \qquad y'' + p \cdot y' + q \cdot y = 0,$$

which possess altogether algebraical solutions. In fact, this problem is solved on the basis of those developments, which we have already brought forward in the third chapter, respecting linear differential equations of the third order, and this so simply that it would seem wrong to pass it over here.

We first replace the differential equation (7), as we did in the third chapter (§ 7), by that differential equation of the third order:

$$(8) \qquad [\eta]_Z = \frac{\eta'''}{\eta'} - \frac{3}{2}\left(\frac{\eta''}{\eta'}\right)^2 = 2q - \frac{1}{z}p^2 - p' = r(Z),$$

which is satisfied by the quotient η of two arbitrary solutions

y_1, y_2, of (7). Evidently η is algebraical, if y_1 and y_2 are so. Let us now recall the formulæ (Chapter III, Equation 33)):

$$(9) \qquad y_1 = \eta \cdot y_2, \quad y_2 = \sqrt{\frac{k}{\eta'}} \cdot e^{-\frac{1}{2}\int p dZ},$$

then we see that we can reverse then, and only then, when $\int pdZ$ is the logarithm of an algebraic function.* *This is a first condition to be laid upon the coefficient* p. Supposing this fulfilled in what follows, we can altogether disregard the equations (7), and now have further the problem, to establish all algebraically integrable equations (8), where $r(Z)$ denotes an unknown, but in any case *rational*, function. We then treat this problem by first presenting all algebraical *integral equations* which, on differentiation, lead to differential equations of the third order of the kind we seek; the establishment of the differential equations themselves will then follow from this very readily.

The function $\eta(Z)$, as an algebraical function of Z, will possess a finite number of branch-points in the plane Z; we will connect these by a network of barriers in such a way that the plane Z acquires a simply connected boundary curve. In the plane so partitioned we then construct, to begin with, a primary function-branch η_0, of necessity everywhere single-valued, which satisfies the differential equation (8). The most general function-branch which satisfies (8) will, by the fundamental property of the differential expression $[\eta]_Z$, be a linear function of this η_0. Hence, as often as we carry η_0 across one of these barriers, it experiences a linear substitution (of course, only dependent on the barrier. We therefore obtain for our η_0 a *group* of linear substitutions, if we traverse all the possible barriers in any kind of combination or repetition. Now we require that η_0 should depend *algebraically* on Z. Hence the number of the function-branches which arise from η_0 by crossing the barriers, and thus also the number of the linear substitutions which η_0 experiences, must be *finite*. We therefore come back at once to the statement of the question given in the preceding paragraph, and can express the result of it forthwith in the following form:

* Since p is to be rational, we can equally well say that $\int pdZ$ is to be the logarithm of a *rational* function.

If η_0 is to be algebraical in Z, there is a linear function z of η_0, for which either z^N, or one of the other fundamental functions $c\frac{F_2^{\nu_2}}{F_3^{\nu_3}}$, remains unaltered when any barriers of the Z-plane are crossed.

This z is of course itself a solution of (8). On the other hand, the expression which remains unaltered, since it is to be an algebraical function of Z, must be a rational function of Z. Hence we have:

If the equation (8) is to be algebraically integrable, the integral equation must, with a proper choice of the particular solution z, be of one of the five forms:

$$(10)\qquad z^N = R(Z), \qquad c\frac{F_2^{\nu_2}(z)}{F_3^{\nu_3}(z)} = R(Z),$$

understanding by $R(Z)$ a rational function of Z.

We now derive conversely, from any one of the equations (10), the value of $[z]_Z$. To this end we write for a moment:

$$z^N = Z_1, \quad c\frac{F_2^{\nu_2}(z)}{F_3^{\nu_3}(z)} = Z_1$$

respectively; then by our previous investigations:

$$[z]_{Z_1} = \frac{N^2-1}{2N^2 . Z_1^2}, \text{ or } = \frac{\nu_1^2-1}{2\nu_1^2(Z_1-1)^2} + \frac{\nu_2^2-1}{2\nu_2^2 . Z_1^2} + \frac{\frac{1}{\nu_1^2}+\frac{1}{\nu_2^2}-\frac{1}{\nu_3^2}-1}{2(Z_1-1)Z_1}$$

respectively. Now we found, on the other hand, in § 6 of the third chapter the general formula:

$$[z]_Z = \left(\frac{dZ_1}{dZ}\right)^2 \cdot [z]_{Z_1} + [Z_1]_Z.$$

On here introducing for Z_1 its value $R(Z)$, we obtain the following differential equations, which $\eta = z$ satisfies:

$$(11)\begin{cases} [\eta]_Z = \dfrac{N^2-1}{2N^2 . R^2} \cdot R'^2 + [R]_Z, \\ [\eta]_Z = R'^2 . \left\{\dfrac{\nu_1^2-1}{2\nu_1^2(R-1^2)} + \dfrac{\nu_2^2-1}{2\nu_2^2 . R^2} + \dfrac{\frac{1}{\nu_1^2}+\frac{1}{\nu_2^2}-\frac{1}{\nu_3^2}-1}{2(R-1)R}\right\} + [R]_Z. \end{cases}$$

These differential equations are evidently included under the formulæ (8), inasmuch as in them also a rational function of

Z occurs on the right-hand side. *Hence we conclude that the rational function* $R(z)$ *introduced in the formulæ* (10) *may be absolutely any rational function, and that, in this sense, the equations* (11), *to which the equations* (10) *correspond as particular solutions, are the most general differential equations of those we seek.* Thus is the problem, which we formulated at the beginning of this paragraph, fully solved.*

§ 4. Finite Groups of Linear Substitutions for a Greater Number of Variables.

Turning now to the first of the generalisations proposed in § 1, my intention is not to communicate† examples of finite groups of linear substitutions for a greater number of variables, or otherwise to enter into particulars with regard to these groups. I am rather concerned just to explain, on general lines, how *fundamental problems* admit of being formulated corresponding to any such group.

Let our group be first written in the *homogeneous* form. Then certain *integral functions* of the variables $z_0, z_1, \ldots z_{n-1}$ (forms) will be given, which are not altered for the substitutions of the group. We will seek to establish the *complete system* of these forms, *i.e.*, those forms:

$$F_1, F_2 \ldots\ldots F_\nu,$$

* After *Herr Schwarz* in the oft-quoted memoir in Bd. 75 of Borchardt's Journal (1872) had investigated for the differential equation of the hypergeometrical series all the cases which are algebraically integrable, the question of the most general algebraically integrable linear differential equation of the second order with rational coefficients was attacked by *Herr Fuchs* in the essays just mentioned (1875–78). In connection with the first of his communications I gave (Sitzber. der Erlanger Soc., June 1876; see also Math. Ann. Bd. xi) the simple result now deduced in the text. *Cf.* here also Brioschi: "La théorie des formes dans l'intégration des éq. diff. lin. du second ordre," in Bd. xi of Math. Ann. (1876), also my second essay, "Ueber lineare Diff. gleichungen," in Bd. xii of the same (1877). Further questions also related to linear differential equations of the second order are dealt with upon the same method by *M. Picard* ("Sur certaines éq. diff. lin.," Comptes Rendus de l'Acad. des Sciences, t. 90 (1880). *Halphen's* researches on differential equations of a higher order have been just mentioned.

† For such examples see the already-mentioned works of *C. Jordan*, also my essays in Mathematische Annalen, Bd. iv, p. 346, &c., Bd. xv, p. 251, &c. A special case will have to be dealt with in the third chapter of the following part.

by means of which all other entirely invariant forms admit of being expressed as integral functions. Among them certain identities must subsist which we compute collectively. We suppose now the numerical values of the F's given, in agreement with these identities, but otherwise arbitrarily. Then we have the *form-problem* which corresponds to our group if we attempt the calculation of the corresponding $z_0, z_1, \ldots z_{n-1}$ from these numerical values. The form-problem has as many solutions as the given group contains operations, and all these systems of solution proceed from any one of themselves by means of the operations of the group.

By the side of this form-theory conception that other presents itself which only takes into consideration the *ratios* of the $z_0, z_1, \ldots z_{n-1}$, and therefore works with $(n-1)$ *absolute* variables and *fractional* linear substitutions. Instead of the forms $F_1, F_2, \ldots$ we shall now have to take into consideration certain *rational functions*, $Z_1, Z_2, \ldots$ which are composed of the F's—or of such forms as only change by a factor for the homogeneous substitutions—as quotients of null dimensions, and which must be so chosen that all other rational functions, which remain unaltered, must be composed rationally of them. In order, then, to seek all the identities subsisting between these Z's, let us suppose the numerical values of the Z's given, in agreement with these identities, but otherwise arbitrarily. We require to compute from them the ratios of the z's. Then we have that which I will generally designate as the *equation-system* belonging to the group. The equation-system has, in relation to the non-homogeneous substitutions of the group, properties quite similar to those which the form-problem has in relation to the homogeneous.

Both problems—the form-problem and the equation-system—can then be assigned a place in the scheme of the *Galois* theory. We might evidently say, making use of the general mode of expression of § 6 in the preceding chapter, that both are their own Galois resolvents. Moreover, it is evident that the solution of the form-problem carries with it that of the equation-system, while the converse need not necessarily be the case.

We will not linger too long over such generalities. On the other hand, we may convince ourselves that in a certain sense the entire theory of equations, commonly so called, will be spanned by these enunciations. If we are concerned with the solution of an equation of the nth degree $f(x)=0$, we can regard it as being the same as if a form-problem for the n variables $x_0, x_1, \ldots x_{n-1}$ (*i.e.*, the roots of the equation) were proposed to us. The group of the corresponding linear substitutions will be simply formed by those permutations of the x's which make up the "Galois group" of the equation; the forms F coincide with the complete system of those integral functions of the x's which, in the sense of the Galois theory, figure as "rationally known." With these remarks, nothing, of course, is primarily altered in the substance of the theory of equations. But the theorems to be developed in it acquire a new arrangement. Those appear as the simplest problems which relate to groups of binary linear substitutions, *i.e.*, just those problems with which we have been dealing in the past chapter. There follow, further, the ternary problems, &c., &c.*

§ 5. Preliminary Glance at the Theory of Equations of the Fifth Degree, and Formulation of a General Algebraical Problem.

The short remarks of the preceding paragraph suffice to exhibit the developments of the second Part here following under that aspect which I suggested in § 1 of the present chapter. Our object will be, in our second Part, to reduce the solution of the general equation of the fifth degree, after adjunction of the square root of the discriminant, to the solution of the icosahedral equation. We have here, in the equation of the fifth degree, as a consequence of the conception just expounded, a *form-problem* with 5 variables, and a group of 60 linear substitutions, before us. On the other side we have, in the icosahedral equation, an *equation-system* (if this expression is allowed for the case of only one variable) also with a group of 60 substitutions, and this is a group which, as we know, is simply isomorphic with the group of the given equation of the

* The conception thus formulated was founded essentially in my essays in Bd. iv of the Mathematische Annalen (1871): "Ueber eine geometrische Repräsentation der Resolventen algebraischer Gleichungen." See, too, the memoir in Bd. xv of the same, to be presently mentioned.

fifth degree. While dealing with this particular question—and this with geometrical reflexions which, under this form, only find a place in this connection—we therefore obtain a contribution to the general problem: to investigate thoroughly *how far it is possible to reduce form-problems or equation-systems, with respectively isomorphic groups, to one another*. By isomorphism we need not necessarily, of course, understand simple isomorphism.

The formulation of this problem has a certain importance, for we obtain thereby at the same time a general programme for the further development of the theory of equations. Among the form-problems and equation-systems with isomorphic groups, we have already above described as the simplest that which possesses the smallest number of variables. If, therefore, any equation $f(x)=0$ is given, we will first investigate what is the smallest number of variables with which we can construct a group of linear substitutions which is isomorphic with the Galois group of $f(x)=0$. Then we shall establish the form-problem or the equation-system which appertains to this group, and then seek to reduce the solution of $f(x)=0$ to this form-problem or equation-system, as the case may be.

The limits of the matter, within which I should like to keep in the present exposition, make it impossible for me to enter more minutely into the aspect thus described. I will merely, while considering equations of the fifth degree, show cursorily how we can treat equations of the third and fourth degree in an analogous sense, by combining the former with the dihedral equation of the sixth degree, and the latter with the octahedral equation (or, if the square root of the discriminant is adjoined, with the tetrahedral equation). I would the more earnestly commend the consideration here of an essay (in Bd. xv of the Mathematische Annalen *) and the associated researches of *Herr Gordan*.† There the principles under consideration are so far developed that a satisfactory theory of the equations of the seventh and eighth degrees with a Galois group of 168 permutations can be established, a theory which appears as a

* "Ueber die Auflösung gewisser Gleichungen von siebenten und achten Grade" (1879).

† See especially "Ueber Gleichungen siebenten Grades mit einer Gruppe von 168 Substitutionen" in Bd. xx of the Mathematische Annalen (1882).

natural extension of the theory of equations of the fifth degree given in the following pages.*

§ 6. Infinite Groups of Linear Substitutions of a Variable.

We now pass to the second generalisation of the previous statement of the question. We shall not alter the number of the variables, but the number of the substitutions, inasmuch as we start from *infinite* groups instead of finite groups. Neglecting the form-theory standpoint, I will here only make mention, in a function-theory form, of the most simple examples of all.† In the place of the rational functions of z (which remain unaltered for the groups of a finite number of substitutions), we have then transcendent but one-valued functions.

Let us first consider in this sense the *simply periodic* and the *trigonometric* functions.

A periodic function of z satisfies the fundamental equation:

(12) $$f(z+ma)=f(z),$$

where m can denote any positive or negative integral number. We have here, therefore, the group of substitutions:

(13) $$z'=z+ma,$$

in relation to which the z-plane is decomposed in the well-known manner into an infinite number of "equivalent" parallel strips, which are "fundamental domains" for the group in the sense before explained. The simplest periodic function:

* If we wished to treat equations of the sixth degree in an analogous sense, it would be necessary, after adjunction of the square root of the discriminant, to start from that group of 360 linear transformations of space which I have established in Bd. iv of the Math. Ann., l.c., and to which latterly Signor Veronese has returned from the side of geometry ("Sui gruppi P_{360}, Π_{360} della figura di sei complessi lineari di rette," &c , Annali di Matematica, ser. 2, t. xi, 1883).

† It would follow that we should make special mention of the *doubly periodic functions* also. But these have a somewhat more complicated character than the other examples. For in their case there is no individual fundamental function by means of which all the rest express themselves rationally; we must rather start from *two* functions, Z_1, Z_2 (between which an algebraical relation of deficiency $p=1$ exists).

(14) $$e^{\frac{2i\pi z}{a}} = Z,$$

assumes within such a strip every value once; in consequence of which all other periodic functions, which arrive at every value a finite number of times in an individual parallel strip, express themselves rationally in terms of Z. We see that this Z plays, with regard to the group (13), the same rôle as formerly the rational fundamental function denoted by the same symbol played in the case of the finite groups. We can also, as in the case of the finite groups, speak of an "equation" which appertains to our group. This is simply formula (14) conceived in the sense of our requiring to compute z from a given Z. Let us consider here that we can look upon $e^{\frac{2i\pi z}{a}}$ as the limiting case of a power with an increasing exponent, and accordingly (14) as the limiting case of a binomial equation. To this end it suffices to recall the well-known definition:

(15) $$e^x = \left(1 + \frac{x}{n}\right)^n_{\lim n = \infty}.$$

We find the transition to the *trigonometric* functions on combining with (13) the new substitution:

(16) $$z' = -z,$$

and thus doubling the number of the substitutions to be taken into consideration. To obtain appropriate fundamental domains appertaining to the new group, let us draw the straight line which contains the point $z = ma$, and decompose, by means of it, each of the parallel strips hitherto considered into two parts. In place of the fundamental function (14) the following now occurs:

(17) $$e^{\frac{2i\pi z}{a}} + e^{\frac{-2i\pi z}{a}} = 2 \cos \frac{2\pi z}{a};$$

our "equation" requires us therefore to compute, from the value of the cosine, the value of the argument. This equation is also a limiting case of the former one. Namely, let us first write the dihedral equation:

$$\frac{(z_1^n - z_2^n)^2}{4z_1^n z_2^n} = -Z,$$

in the following form:

$$z^n + \frac{1}{z^n} = -4Z + 2\,;$$

let us then substitute $1+\frac{x}{n}$ for z, and allow n to increase beyond all limits; then the left side:

$$\left(1+\frac{x}{n}\right)^n + \left(1+\frac{x}{n}\right)^{-n},$$

is transformed into 2 cos ix.

Over and above these familiar examples, let us now consider further the *elliptic modular functions*, and certain other functions related to them which *Herr Schwarz* was the first to take into consideration in his oft-quoted memoir on the hypergeometrical series (in Borchardt's Journal, Bd. 75, 1872). In § 8 of our third chapter we have, as was before explained, established for the root z of the dihedral, tetrahedral, octahedral, and icosahedral equations in common, the differential equation of the third order:

$$(18) \qquad [z]_Z = \frac{\nu_1^2-1}{2\nu_1^2(Z-1)^2} + \frac{\nu_2^2-1}{2\nu_2^2 \cdot Z^2} + \frac{\frac{1}{\nu_1^2}+\frac{1}{\nu_2^2}-\frac{1}{\nu_3^2}-1}{2(Z-1)Z},$$

where for ν_1, ν_2, ν_3 respectively the numerical values in the following oft-used table were to be introduced:

	ν_1	ν_2	ν_3
Dihedron . .	2	2	n
Tetrahedron . .	2	3	3
Octahedron . .	2	3	4
Icosahedron . .	2	3	5

and indeed these are, as we just showed (in § 2), the only numerical values for which $\frac{1}{\nu_1}+\frac{1}{\nu_2}+\frac{1}{\nu_3}$ is >1. *The functions of Herr Schwarz arise on inserting in* (18) *for* ν_1, ν_2, ν_3, *any other three integers* (whereupon $\frac{1}{\nu_1}+\frac{1}{\nu_2}+\frac{1}{\nu_3}$ will be $\overline{\lessdot}1$).

In order to give a representation of the march of these functions, let us remark the following. In the third chapter we have seen that, in virtue of our fundamental equations, the half-plane Z will be represented on spherical triangles of the z-sphere, the angles of which are respectively $\frac{\pi}{\nu_1}, \frac{\pi}{\nu_2}, \frac{\pi}{\nu_3}$. Just the same takes place in the case of the functions we are now speaking of, as soon as we have fixed upon the particular solution of (18), which we wish to take into consideration, and then develop this analytically. But while, corresponding to the algebraical character of the fundamental equations, a *finite* number of spherical triangles then sufficed to cover the z-sphere, now an infinite number of such triangles (no one of which infringes on another) are placed side by side. We must here distinguish when $\frac{1}{\nu_1}+\frac{1}{\nu_2}+\frac{1}{\nu_3}$ is $=1$ or <1. In the first case, all the spherical sides which bound the triangles pass, when produced, through a fixed point on the z-sphere, and we approach nearer and nearer to this fixed point as we multiply the triangles in succession, without, however, actually reaching it. The function Z has a finite value everywhere, except at this point.

In the other case, the bounding spherical lines have a common orthogonal circle, and this circle forms the limit which we approach, by increasing the spherical triangles, as near as we like, without, however, crossing it. Hence the function $Z(z)$ exists only on one side of the orthogonal circle; the orthogonal circle is for us what is described as *a natural boundary*.* As regards the corresponding group of linear substitutions, let us consider the spherical triangles in question alternately shaded and non-shaded. The group then consists of all linear substitutions of z which change a shaded triangle into another shaded triangle (or a non-shaded into a non-shaded triangle).

Amongst the functions thus introduced, the *elliptic modular functions* now form (to limit ourselves to the simplest kind) a special case, the case $\nu_1=2$, $\nu_2=3$, $\nu_3=\infty$. The spherical triangle of the z-sphere has then, corresponding to the value of ν_3, an angle equal to zero. If we allow the limiting circle,

* *Cf.* throughout the memoir of Schwarz above cited, in which, moreover, appropriate figures are given.

which $Z(z)$ possesses on the z-sphere, to coincide with the meridian of real numbers, we are able to ensure that the totality of the corresponding linear substitutions is given by those *integral, real* substitutions:

$$z' = \frac{\alpha z + \beta}{\gamma z + \delta},$$

whose determinant $(\alpha\delta - \beta\gamma)$ *is* $=1$. Let g_2, g_3 be the invariants of a binary biquadratic form $F(x_1, x_2)$ (see § 11 of the second chapter), then it is known that $\Delta = g_2^3 - 27g_3^2$ is the corresponding discriminant. Now put the Z in question equal to the absolute invariant $\frac{g_2^3}{\Delta}$. *Then the* $z(Z)$ *is nothing else than the ratio of two primitive periods of the elliptic integral:*

$$\int \frac{x_2 dx_1 - x_1 dx_2}{\sqrt{F(x_1 x_2)}},$$

therefore the $\frac{iK'}{K}$ *of the Jacobian notation.**

It is impossible to enter here more minutely into the various relations thus touched upon. We will only bring forward this remark, that, in virtue of the conception developed, the elliptic modular functions appear, just in the same way as the exponential function and the cosine, as the last term of a series of infinitely many analogously constructed functions. Put in formula (18) ν_1 throughout equal to 2, ν_3 equal to 3, and then let ν_2, beginning with 2, assume successively all integral positive values. Then we have for $\nu_3 = 2$ a case of the dihedron† (only that ν_2 is taken $> \nu_3$, whereas we have usually elsewhere arranged the ν's in the order of the magnitudes), for $\nu_3 = 3, 4, 5$, the tetrahedron, octahedron, and icosahedron, in order; then

* See *Dedekind* in Borchardt's Journal, Bd. 83 (1877), also my essay "Ueber die Transformationen der elliptischen, Functionen,"&c., in Bd. xiv of the Math. Ann. (1878). Anyone specially interested in this theory should consult particularly the memoir of Herr Hurwitz in Bd. 18 of Mathematische Annalen (1881) "Grundlagen einer independenten Theorie der elliptischen modulfunctionen," etc.).

† It is the same case to which, as explained in § 7 of the second chapter, the calculation of the double ratio of four points leads, or of the modulus of the elliptic functions, and which, moreover, will form a starting-point in the sequel for the solution of equations of the fifth degree.

for greater values of ν_3 an infinite series of transcendent functions, whose termination for $\nu_3 = \infty$ is formed by the elliptic modular functions.

§ 7. Solution of the Tetrahedral, Octahedral, and Icosahedral Equations by Elliptic Modular Functions.

Short as the preceding suggestions are, they suffice to make intelligible how it comes to pass that we can solve the equations of the tetrahedron, octahedron, and icosahedron (or indeed the special case of the dihedral equation just mentioned) by means of elliptic modular functions. Let us first consider the *logarithmic* solution of the binomial equation:

$$z^n = Z,$$

or what is quite analogous, the *trigonometric* solution of the dihedral equation:

$$z^n + z^{-n} = -4Z + 2.$$

Both equations admit of being regarded as a limiting case of a trivial algebraical solution, which consists in first calculating ζ from the equation:

$$\zeta^{mn} = Z, \text{ or } \zeta^{mn} + \zeta^{-mn} = -4Z + 2,$$

understanding by m any positive integral number, and then finding z equal to a rational function of ζ:

$$z = \zeta^m.$$

The transcendent solution proceeds from this on taking $m = \infty$, whereupon ζ^{mn} is transformed into e^{ζ} just in the manner described, $\zeta^{mn} + \zeta^{-mn}$ into $2 \cos i\zeta$, while z will become $e^{\frac{\zeta}{n}}$.

The case is precisely the same now with the representation of our fundamental irrationalities by means of elliptic functions. We convince ourselves, first, that each of the Schwarzian functions ν_1, ν_2, ν_3 admits of being represented uniquely by means of every other ν_1', ν_2', ν_3', the exponents of which are integral multiples of the original ν_1, ν_2, ν_3. In particular, therefore, if we limit ourselves to that series of functions for which $\nu_1 = 2$, $\nu_2 = 3$, the only condition necessary for a single-valued representation will be that ν_3' is divisible by ν_3. But this is always the case if $\nu_3' = \infty$. *All functions of our series, therefore, admit of a single-valued representation in terms of the elliptic modular functions, and it is just this which is described as a solution of the equations in question by the help of the elliptic modular functions.*

I communicate here, without proof, the simplest formulæ which present themselves in this direction for the tetrahedron, octahedron, and icosahedron.* We write the three several fundamental equations, as we have always done, in the following manner:

$$\frac{\Phi^3}{\Psi^3} = Z, \quad \frac{W^3}{108 f^4} = Z, \quad \frac{H^3}{1728 f^5} = Z.$$

Then let Z, as just now, be the absolute invariant $\frac{g_2^{\,3}}{\Delta}$ of an elliptic integral of the first kind, $\frac{iK'}{K}$ the ratio of its periods, $q = e^{-\frac{K'}{K}\cdot\pi}$. Then we have first, for the root of the *octahedral equation*, the simple formula:

$$(19) \qquad z = q^{\frac{1}{2}} \cdot \frac{\sum_{-\infty}^{+\infty} q^{2\kappa^2 + 2\kappa}}{\sum_{-\infty}^{+\infty} q^{2\kappa^2}};$$

this arises from the known equation:

$$\sqrt{k} = \frac{\theta_2(o, q)}{\theta_2(o, q)},$$

on introducing q^2, instead of q, on the right-hand side.†

* Compare Bd. xiv of the Math. Ann., pp. 157, 158; also the essay of Signor Bianchi: "Ueber die Normalformen dritter und fünfter Stufe des ellip. Integ. erster Gatt."; and my own note: "Ueber gewisse Theilwerthe der Θ-Functionen" in Bd. xvii, *ibid.* (1880–81).

† This corresponds to the remark which we made above (p. 48 of the text) concerning certain researches of Abel's. In order to thoroughly understand the connection of what is to follow in this direction, let us compute for the biquadratic form $(1 - x^2)(1 - k^2x^2)$ the absolute invariant $\frac{g_2^{\,3}}{\Delta}$. We then obtain:

$$\frac{(1 + 14k^2 + k^3)^3}{108k^2(1 - k^2)^4},$$

and on inserting here for $\sqrt{k}$ the letter z, we have exactly the left side of the octahedral equation; the symbols θ_2, θ_3, as also θ_1, which I employ in the text, are the well-known Jacobian ones.

We find, further, for the *icosahedral irrationality:*

$$(20)\qquad z = q^{\frac{2}{5}} \cdot \frac{\sum\limits_{-\infty}^{+\infty}(-1)^{\kappa}.\, q^{5\kappa^2-3\kappa}}{\sum\limits_{-\infty}^{+\infty}(-1)^{\kappa}.\, q^{5\kappa^2-\kappa}} = q^{\frac{2}{5}} \cdot \frac{\theta_1\left(\frac{2iK'\pi}{K}, q^5\right)}{\theta_1\left(\frac{iK'\pi}{K}, q^5\right)},$$

an expression, therefore, which coincides with

$$\frac{q^{\frac{2}{5}}}{1+q^2}$$

when the term involving q^{10} is disregarded.

The solution of the *tetrahedral equation* takes a rather more complicated form. We will in this case first replace the z hitherto used by a linear function of z, which vanishes at the summits of $\Psi = 0$, and becomes infinite at the opposite summits of $\Phi = 0$. In this sense we write:

$$(21)\qquad z = (1+i)\frac{-\xi + (\sqrt{3}+1)}{(\sqrt{3}+1)\xi + 2}.$$

For the ξ thus defined we have then, first, the equation:

$$(21\text{a})\qquad Z = \frac{g_2^{\,3}}{\Delta} = 64\,\frac{(\xi^3-1)^3}{\xi^3\,(\xi^3+8)^3},$$

and, further, the transcendent solution:

$$(21\text{b})\qquad \xi = -6q^{\frac{2}{3}}.\,\frac{\sum\limits_{0}^{\infty}(-1)^{\kappa}\,(2\kappa+1)\;q^{3+3}}{\sum\limits_{-\infty}^{+\infty}(-1)^{\kappa}\,(6\kappa+1)\;q^{3\kappa^2+\kappa}}.$$

We have thus for our three equations severally determined one root; we obtain the remaining corresponding roots if we substitute in $q = e^{-\frac{K'}{K}\cdot\pi}$ for $\frac{iK'}{K}$ the infinite number of values:

$$\frac{\alpha\cdot\frac{iK'}{K}+\beta}{\gamma\cdot\frac{iK'}{K}+\delta},$$

where α, β, γ, δ are real integers of determinant 1. *Here all such systems* $\alpha, \beta, \gamma, \delta$ *as coincide for modulus* ν_3, *or can be brought* *into coincidence by means of a uniform change of sign, always give rise to the same root.**

§ 8. FORMULÆ FOR THE DIRECT SOLUTION OF THE SIMPLEST RESOLVENT OF THE SIXTH DEGREE FOR THE ICOSAHEDRON.

In accordance with the particular significance which we attach to the icosahedral equation, the second of the formulæ (19)—(21) of course has most interest for us. We have already explained that the simplest resolvent of the sixth degree which the icosahedral equation possesses has been placed by *Herr Kronecker* in direct relation with the modular equation of the sixth order for a transformation of the fifth order of the elliptic functions (see § 15 of the preceding chapter). The formula in question has since been considerably simplified by *Herr Kiepert* and myself by the introduction of the rational invariants.† Since, in our researches on equations of the fifth degree, much regard will be paid to this very formula, it may also be communicated here, the proof being left out, and the symbols used elsewhere being adapted.

We have in § 15 of the preceding chapter (formula (46)) given the following form to the resolvent alluded to:

$$\zeta^6 - 10Z\,.\,\zeta^3 + 12Z^2\,.\,\zeta + 5Z^2 = 0.$$

Now let g_2, Δ be the invariants, already so denoted, of an elliptic integral, and let Z be taken $=\frac{g_2^3}{\Delta}$. Further, let Δ' be that value which is derived from Δ by any transformation of the fifth order. *Then the root of our resolvent is simply:*

$$(22) \qquad \zeta = -\frac{g_2 \sqrt[12]{\Delta'}}{\sqrt[12]{\Delta^5}}.$$

* ν_3 is $=3$ for the tetrahedron, $=4$ for the octahedron, $=5$ for the icosahedron. For the special dihedral equation appertaining hereto, the same theorem would hold good for $\nu_3=2$. Compare for this, Mathematische Annalen, Bd. xiv, pp. 153, 156.

† *Cf.* Bd. xiv of the Math. Ann., p. 147; also Bd. xv, p. 86 (1878); and further, *Kiepert:* "Auflösung der Gleichungen 5. Grades," and "Zur Transformations-theorie der elliptischen Functionen" (Borchardt's Journal, Bd. 87, 1878-79); and, finally, the memoir of *Hurwitz* just mentioned.

If we like to express everything here in terms of K, iK', and q respectively, and thereby to derive at the same time from each other the six different roots (22), we have first to insert the following values for g_2 and $\sqrt[12]{\Delta}$:

$$(23) \begin{cases} g_2 = \left(\frac{\pi}{K}\right)^4 \cdot \left\{ \frac{1}{12} + 20 \left(\frac{q^2}{1-q^2} + \frac{2^3 q^4}{1-q^4} + \frac{3^3 q^6}{1-q^6} + \cdots \right) \right\}, \\ \sqrt[12]{\Delta} = \left(\frac{\pi}{K}\right) \cdot q^{\frac{1}{6}} \cdot \prod_{\kappa=1}^{\infty} (1-q^{2\kappa})^2, \end{cases}$$

and then for $\sqrt[12]{\Delta'}$ to put the following six values respectively:

$$(24) \begin{cases} \sqrt[12]{\Delta'_\infty} = \left(\frac{5\pi}{K}\right) \cdot q^{\frac{5}{6}} \cdot \prod_{\kappa=1}^{\infty} (1-q^{10\kappa})^2, \\ \sqrt[12]{\Delta'_\nu} = \left(\frac{\pi}{K}\right) \cdot \epsilon^{2\nu} q^{\frac{1}{30}} \cdot \prod_{\pi=1}^{\infty} \left(1 - \epsilon^{4\kappa\nu} \cdot q^{\frac{2\kappa}{5}}\right)^2, \end{cases}$$

where $\nu = 0, 1, 2, 3, 4$, and ϵ is to be taken $= e^{\frac{2i\pi}{5}}$. The indices ∞, ν are here chosen exactly as in § 15 of the previous chapter. The formula (23) can at the same time be used to complete the data of the preceding paragraph; namely, the absolute invariant of the elliptic integral is given by them in the form:

$$(25) \qquad \frac{g_2^3}{\Delta} = \frac{1}{1728 q^2} \cdot \frac{\left(1 + 240 \sum_1^{\infty} \kappa^3 \cdot \frac{q^{2\kappa}}{1-q^{2\kappa}}\right)^3}{\prod_1^{\infty} (1-q^{2\kappa})^{24}}.$$

§ 9. Significance of the Transcendental Solutions.

The significance of the transcendental solution with which we have now become acquainted is primarily a purely practical one. Logarithms, trigonometric functions, and elliptic modular functions have been long tabulated, in consideration of the importance which they possess in other fields of analysis. By reducing the solution of our equations to the said transcendent functions, we make these tables available for use, and avoid the tedious calculation which would be necessary in carrying out the method of solution by means of hypergeometrical series given in Chapter III.*

But there is a deeper conception of the transcendental solutions, by which the latter lose the foreign aspect which they seem to have in the midst of our other investigations, and indeed are seen to be in intimate relation with them.

Let us consider, in order to fix our ideas, say, the solution of the icosahedral equation as it is furnished by (20). As often as we subject $\frac{iK'}{K}$ to one of the infinitely many corresponding linear integral substitutions, the z, in virtue of this formula, experiences one of the 60 linear icosahedral substitutions. *The group of the substitutions of* $\frac{iK'}{K}$ *therefore appears isomorphously related to the group of the* 60 *icosahedral substitutions.* The isomorphism is only, if we may so express it, one of "infinitely high" merihedry: to the individual substitution of $\frac{iK'}{K}$ corresponds one, and only one, substitution of z, while to every substitution of z correspond infinitely many substitutions of $\frac{iK'}{K}$. Let us now recall the considerations of § 5. Limiting ourselves there to *finite* groups of linear substitutions, we required to bring into connection with one another such equation-systems generally (or form-problems) as are related to isomorphic groups. We now extend this problem to *infinite* groups of linear substitutions, and recognise *that our transcendent solutions realise special cases of the problem so generalised.* We have obtained these solutions by making use of the theories, developed in other quarters, of certain transcendent functions. This is evidently a process which, in connection with our present considerations, is not theoretically satisfactory. *We rather require a general treatment by means of which the developments given in* § 5, *as well as our present transcendent solutions, will be furnished.* Our reflections thus lead to a comprehensive problem, which will embrace the theory of equations of a higher

* The unfortunate circumstance here arises, as regards the elliptic modular functions, that *Legendre's* tables for the calculation of the elliptic integrals have not yet been formed in a way which would correspond with *Weierstrass's* theory of elliptic function.

degree, as well as the law of construction of the θ-function. In proposing this problem, however, we have reached the limit, as in § 5, which bounds our present exposition, and which we may not pass.*

* I will, however, not omit to call attention here to certain developments of M. Poincaré's (on the general function which he denotes by Z) which behave just in the way here alluded to; see Mathematische Annalen, Bd. 19, pp. 562, 563 (1881).

I have, further, to append here the following quotations, which resemble one another in relating to works which, with greater or less completeness, the theories expounded in our first Part are connectedly dealt with: (1) *Puchta*, "Das Oktaeder und die Gleichung vierten Grades," Denkschriften der Wiener Akademie, math.-phys. Kl., Bd. 91 (1879). This work might also be consulted throughout the following Part when we are concerned with the solution of equations of the fourth degree (by means of the octahedral equation).—(2) *Cayley*, "On the Schwarzian Derivative and the Polyhedral Functions," Transactions of the Cambridge Philosophical Society, vol. xiii (1880). By the "Schwarzian derivative" is there understood the differential expression of the third order, which we established in § 6 of the third chapter.—(3) *Wassilieff*, "Ueber die rationalen Functionen, welche den doppeltperiodischen analog sind," Kasan (1880) (Russian). Herr Wassilieff there makes the interesting remark that *Hamilton* had already considered the group of the icosahedral rotations with reference to their generation from two operations. ("Memorandum respecting a New System of Non-Commutative Roots of Unity," Philosophical Magazine, 1856.)

PART II

THEORY OF EQUATIONS OF THE FIFTH DEGREE

CHAPTER I

THE HISTORICAL DEVELOPMENT OF THE THEORY OF EQUATIONS OF THE FIFTH DEGREE

§ 1. Definition of our First Problem.

THE considerations of the preceding Part have given us a determinate problem with regard to equations of the fifth degree: we wished to try to effect this solution by the help of the icosahedron. Now it would not be difficult to put the results, which I have to develop in this connection, as such in the foreground, and derive them in a deductive form. I prefer, however, to avail myself of the inductive method here also, and this in such a way that, on the one hand, I pay regard to the historical development of equations of the fifth degree, and, on the other hand, make free use of geometrical constructions. I hope in this way to unfold to the reader not only the accuracy of definite results, but also the process of thought which led to them.

In accordance with what has been said, our first task must, in any case, be that of giving an account and review of works hitherto published which are concerned with the solution of equations of the fifth degree, so far as these works will be used in the sequel. I shall here, for the sake of brevity, leave on one side all such developments as we shall not be immediately concerned with, however weighty and essential these may appear from more general points of view. To these belong, above all, the proofs of *Ruffini* and *Abel*, by which it is established that a solution of the general equation of the fifth degree by extracting a finite number of roots is impossible; and the parallel works, likewise set on foot by *Abel*, in which all special equations of the fifth degree are determined, which differ in this respect from the general equation. To these again belong the efforts of *Hermite* and *Brioschi* to apply the invariant theory of binary forms of the fifth order to the solution of equations of the fifth degree; not that the use of the invariant-theory processes will be altogether dispensed with in the following pages, only that in our case these relate throughout, as in the preceding Part, to such forms as are transformed into themselves by determinate linear substitutions, and not to binary forms of the fifth order. Finally, we leave on one side the question of the reality of the roots of equations of the fifth degree; in particular, therefore, the extended investigations by which *Sylvester* and *Hermite* have made the reality of the roots to depend on the invariants of the binary form of the fifth order.

If we limit our task in the manner here described, there remain two fields of labour which we have to consider. The object of both of them is to study the roots of the general equation of the fifth degree as functions of the coefficients of the equation. Both start by simplifying the functions in question, so that, instead of the five independent coefficients of the equation, a smaller number of independent magnitudes will be introduced. It is only that the means which are employed for this purpose are different; in the one case it is the *transformation of the equations*, in the other it is the *construction of resolvents*.

The method of transformation goes back, as we know, to *Tschirnhaus*.* Let

$$(1) \qquad x^n + Ax^{n-1} + Bx^{n-2} + \ldots . Mx + N = 0$$

be the proposed equation of the nth degree; then Tschirnhaus put:

$$(2) \qquad y = \alpha + \beta x + \gamma x^2 + \ldots . \mu . x^{n-1},$$

whereupon, by elimination of x between (1) and (2), he obtained an equation for y, also of the nth degree, to which he endeavoured to impart special properties by a proper choice of the coefficients $\alpha, \beta, \gamma, \ldots$ We will at once describe the results which have been found by this assumption for the

* "Nova methodus auferendi omnes terminos intermedios ex data æquatione," Acta eruditorum, t. ii, p. 204, &c. (Leipzig, 1683). The title itself shows that Tschirnhaus realised (as Jerrard did later on) the range of his method.

special case of the equation of the fifth degree. Let us first agree that with the y's the x's are also found, at least so long as the equation for the y's, as we of course assume for the equation (1), possesses different roots. For in this case the equations (1) and (2) [in which we now consider y as the unknown magnitude] have only one root x common, and this x can therefore be rationally calculated by known methods.

The method of the construction of resolvents has also long ago been employed for the solution of equations of the fifth degree. Notable in this respect is the year 1771, in which *Lagrange, Malfatti,* and *Vandermonde,* independently of one another, published their closely-related investigations.* However, the results which these attained rather served to point out the existing difficulties than to remove them. It was not till *Herr Kronecker*, in 1858, succeeded in establishing a resolvent of the sixth degree for the equation of the fifth degree, that a real simplification was effected.† We shall have to limit ourselves in our further account, so far as the construction of resolvents is concerned, to the exposition of *Kronecker's* method and the further researches connected with it.

The two fields of labour which we have thus placed beside one another are concerned, *per se*, with purely *algebraical* problems. Howbeit the development of analysis has entailed their both appearing intimately connected with the more extended problem: to effect the solution of equations by the help of proper transcendent functions. We have shown in the last chapter of the preceding Part ‡ that such a use of transcendent functions is primarily of merely practical value, and should not be confounded with the theoretical researches on the theory of equations. However, we must not neglect in the following account to consider the different methods by which the solution

* Lagrange: "Réflexions sur la résolution algébrique des équations," Mémoires de l'Académie de Berlin for 1770–71, or Œuvres, t. iii.

Malfatti: "De æquationibus quadrato-cubicis disquisitio analytica," Atti dell' Accad. dei Fisiocritici di Siena, 1771; also, "Tentativo per la risoluzione delle equazioni di quinto grado," *ibid.*, 1772.

Vandermonde: "Mémoire sur la résolution des équations," Mémoires de l'Académie de Paris, 1771.

† Compare the later references.

‡ I shall in future denote references to the preceding Part by letting the number of the chapter, represented by an arabic number, succeed the roman number I. *Cf.* therefore in this case I, 5, §§ 7, 9.

of equations of the fifth degree has been specially connected with the theory of *elliptic functions.* For it has been just these methods, as we have already suggested, which have led to a clearer conception of the purely algebraical problems also.

For the rest, let it be observed that there is no essential antithesis between the two fields of labour which we are contrasting. If we have succeeded in turning a proposed equation of the nth degree by transformation into another which contains a smaller number of parameters, we can afterwards derive resolvents from the latter, and consider these as peculiarly simple resolvents of the original equation; or conversely, if we have come into possession of a special resolvent of the initial equation by any of these methods, we can return from it, by renewing our formation of resolvents, to an equation of the nth degree, which latter will then admit of being transformed directly from the proposed equation.

§ 2. Elementary Remarks on the Tschirnhausian Transformation—Bring's Form.

In order to compute the equation of the nth degree which the y's of formula (2) satisfy, it is most convenient to construct its coefficients directly, as symmetric functions of the y's, from the symmetric functions of the x's. In this way we recognise at once that *the coefficient of $y^{n-\kappa}$ is an integral homogeneous function of the κth degree of the indeterminate magnitudes $\alpha, \beta, \gamma, \ldots . \nu$.* Hence we have a linear equation with n unknowns to solve, if we wish to expel the term involving y^{n-1} from the transformed equation, and a quadratic equation of the same kind appears in addition if the term involving y^{n-2} is to vanish also. We satisfy both these equations together if we consider $n-2$ of the unknowns as parameters, and determine one of the remaining ones by means of a quadratic equation after eliminating the last unknown. I shall describe an equation in which the terms involving y^{n-1}, y^{n-2}, are wanting as a *canonical equation* for the future. *The Tschirnhausian transformation, therefore, allows us to reduce every equation to a canonical equation with the help of merely a square root.* On the other hand, we meet with difficulties as soon as we require that another term in the equation of the y's shall vanish. In fact, we then come upon elimination-equations of a higher degree, which we do not know how to treat by elementary means. It is here that a more searching investigation has brought to light an important and—for our future exposition—fundamental result. The equation of elimination of which we speak will be of the sixth degree if we require that the terms y^{n-1}, y^{n-2}, y^{n-3} shall vanish simultaneously; *it has been shown that by proper choice of coefficients of transformation for* $n > 4$ *the said equation of the sixth degree can be reduced to an equation of the third degree by the solution of quadratic equations.*

The result thus described is usually ascribed to the English mathematician *Jerrard*, who made it known in the second part of his Mathematical Researches (Bristol and London, 1834, Longman). But it is of much earlier date so far as equations of the fifth degree are concerned. As *Hill* remarked in the Transactions of the Swedish Academy, 1861, it had already been published in 1786 by *E. S. Bring* in a Promotionschrift submitted to the University of Lund.* I should, nevertheless, have retained in the following pages the practice, generally diffused at the present time, of describing it in connection with Jerrard, had not the latter in his works relating to this matter brought forward, amongst some interesting results, a lot of thoroughly false speculations: he believed (just as Tschirnhaus did) that he could remove, by the help of his method, all the intermediate terms, not only from equations of the fifth degree, but equations of any degree, by means of elementary processes, and did not lay aside this view in spite of incisive refutation from the other side.† I shall therefore in future speak of *Bring's* equation. Let us write the canonical equation of the

* The full title runs: "Meletemata quædam mathematica circa transformationem æquationem algebraicarum, quæ preside E. S. Bring . . . modeste subjicit S. G. Sommelius." We might, perhaps, have been led by the title to suppose that Sommelius was the author, but I learn from Herr Bäcklund of Lund that this would certainly be erroneous, inasmuch as the Promotionschriften were then composed entirely by the examiners, and only served the candidates as a substratum for disputation. The principal points of Bring's treatise are reprinted in the communication, already mentioned, of Hill to the Swedish Academy, and again in the Quarterly Journal of Mathematics, vol. vi, 1863 (Harley, "A Contribution to the History," &c.), and, finally, in Grunert's Archiv, t. xli, 1864, pp. 105–112 (with remarks by the editor).

† Jerrard's further publications are found principally in the Philosophical Magazine, vol. vii (1835), vol. xxvi (1845), vol. xxviii (1846), vol. iii (new series) (1852), vols. xxiii, xxiv, xxvi (1862–63), &c., and are, therefore, for the

fifth degree (as it is to be written henceforward) in the following form:

(3) $$y^5 + 5ay^2 + 5by + c = 0.$$

Then it will be to the purpose to retain the coefficient 5 in Bring's form also. On substituting at the same time z for y, for the sake of distinction, we have:

(4) $$z^5 + 5bz + c = 0.$$

Bring's equation still contains, as we see, at first two coefficients. We can, however, at once remove one of them by putting $z = \rho t$ and then suitably determining ρ. *We can, therefore, by a proper Tschirnhausian transformation, effect that the five roots of the equation of the fifth degree shall appear to depend on a single variable magnitude.* This result is more peculiarly important because we are much more completely masters of the functions of a single argument than of those of a larger number of variables. Let us write (4), *e.g.*, as follows (as *Hermite* has done in his researches to be quoted immediately):

$$t^5 - t - A = 0,$$

then it is very easy, on the one hand, to exhibit by Riemann's method how the five roots t depend on A, and, on the other hand, to establish for any values of A appropriate developments in ascending and descending powers which allow the five roots t to be computed to any approximation.

Having thus become acquainted with Bring's *result*, we may postpone a deeper consideration of its basis, and also a criticism on its significance, till later, when we shall have further occasion to do so in connection with our own developments. I also omit to enumerate all the numerous commentaries which the researches of Bring and Jerrard respectively have received in the course of years. One of the first expositions of this method, and, at the same time, the one most widely known, is perhaps that in *Serret's* "Traité d'algèbre supérieure" (1st edition, 1849). *Hermite* has also dealt with Bring's transfor-

most part, later than the report (as lucid as it is voluminous) which *Hamilton* furnished in 1836 for the British Association for the Advancement of Science (Reports of British Association, vol. vi, Bristol). Further, *Cockle* and *Cayley* repeatedly opposed the assertions of Jerrard (Phil. Mag., vols. xvii–xxiv, 1859–62).

mation,* aiming, however, as already observed, at the application of the invariants of binary forms of the fifth degree; we must remark that Hermite has determined the irrationalities necessary for the transformation much more completely than is usually done.

§ 3. Data concerning Elliptic Functions.

The special questions in the theory of elliptic functions on which we must now inform ourselves lie in the region of *the theory of transformation.* With the usual notation let κ be the modulus of an elliptic integral:

$$(5) \qquad \int \frac{dx}{\sqrt{1-x^2 \,.\, 1-\kappa^2 x^2}},$$

λ the modulus which results from a transformation of the nth order, where n is to denote an uneven prime number. Then, according to *Jacobi* † and *Sohnke* ‡ respectively, there exists between $\sqrt[4]{\kappa}=u$ and $\sqrt[4]{\lambda}=v$ an equation of the $(n+1)^{\text{th}}$ degree in each of these quantities, the so-called modular equation:

$$(6) \qquad f(u, v) = 0,$$

which, *e.g.*, for $n=5$, runs as follows:

$$(7) \qquad u^6 - v^6 + 5u^2v^2(u^2 - v^2) + 4uv(1 - u^4v^4) = 0.$$

Here u may be expressed in various ways in terms of $q = e^{-\pi\frac{K'}{K}}$, *e.g.*, as follows:

$$(8) \qquad u = \sqrt{2} \,.\, q^{\frac{1}{8}} \,.\, \frac{\sum q^{2m^2+m}}{\sum q^{m^2}};$$

we obtain the $(n+1)$ values of v, which satisfy the modular equation on inserting in place of $q^{\frac{1}{8}}$ in this formula in order:

$$(9) \qquad q^{\frac{n}{8}}, \quad q^{\frac{1}{8n}}, \quad \alpha q^{\frac{1}{8n}}, \quad \ldots \quad \alpha^{n-1} q^{\frac{1}{8n}},$$

where $\alpha = e^{\frac{2i\pi}{n}}$. *The modular equation therefore gives us an*

* In the comprehensive treatise (which will be often mentioned): "Sur l'équation du cinquième degré," Comptes Rendus, t. lxi, lxii (1865–66). *Cf.* particularly t. lxi, pp. 877, 965, 1073; t. lxii, p. 65.

† "Fundamenta nova theoriæ functionum ellipticarum" (1829).

‡ "Æquationes modulares pro transformatione functionum ellipticarum" (Crelle's Journal, t. xii, 1834).

*example of an equation with one parameter which can be solved by elliptic modular functions.** The parameter is u; we find from it the corresponding q on reversing the formula (8), or calculating the magnitudes K and K' from (5):

$$(10) \qquad K=\int_0^1 \frac{dx}{\sqrt{1-x^2 \cdot 1-\kappa^2x^2}}, \quad K'=\int_0^1 \frac{dx}{\sqrt{1-x^2 \cdot 1-\kappa'^2x^2}},$$

where $\kappa'^2=1-\kappa^2$. The $(n+1)$ roots v are then obtained by means of the substitutions (9).

We now ask whether it is not practicable to effect, by the help of the modular equation, the solution of other equations also. To this end we shall have, above all—in accordance with the explanations which we have given in §§ 1, 4—to determine the *group* of the modular equation. This is what *Galois* himself has already accomplished.† Corresponding to the substitutions (9), *Galois* denotes the roots of the modular equation by the following indices:

$$(11) \qquad v_\infty, v_0, v_1, \ldots v_{n-1}.$$

If we then disregard mere numerical irrationalities,‡ the group of the modular equation is formed of those permutations of the v_ν's which are contained in the following formula:

$$(12) \qquad \nu' \equiv \frac{\alpha\nu+\beta}{\gamma\nu+\delta} \text{ mod. } (n),$$

which we have already considered above in special cases (I, 4, § 15; I, 5, § 7). The coefficients $\alpha, \beta, \gamma, \delta$ are here otherwise arbitrary integers which satisfy the condition $(\alpha\delta-\beta\gamma)\equiv 1$ (mod. n).

We interpret this result specially for $n=5$. The group (12) will then be, as we saw before, simply isomorphic with the group of the 60 icosahedral rotations, *i.e.*, expressed in the abstract, *with the group of even permutations of five things.* We hence

* We have already become acquainted with other examples above, I, 5, §§ 7, 8; since, however, we have here to explain the historical development of the theory, these are for the present not considered.

† See "Œuvres de Galois," Liouville's Journal, t. xi (1846).

‡ According to the researches of Hermite, the single numerical irrationality here coming under consideration is $\sqrt{(-1)^{\frac{n-1}{2}} \cdot n}$. *Cf.* the exposition in C. Jordan's "Traité des substitutions et des équations algébriques," p. 344, &c.

conclude that the modular equation (7) possesses resolvents of the fifth degree, whose discriminant, after adjunction of a numerical irrationality ($\sqrt{5}$, according to *Hermite*), is the square of a rational magnitude. Will it be possible to put the general equation of the fifth degree, after adjunction of the square root of its discriminant, in connection with such a resolvent by means of a Tschirnhausian transformation? Or, conversely, shall we be able, after adjunction of the square root of its discriminant, to establish a resolvent of the sixth degree of the general equation which proceeds from the modular equation (7) by appropriate transformation? These are just the two ways of attacking the solution of equations of the fifth degree by elliptic functions which have been taken in hand and worked out by *Hermite* and *Kronecker* respectively. Before we enter on an account of their results, we have an important addition to make from the theory of elliptic functions.

We mentioned just now the idea of subjecting the modular equation itself to a Tschirnhausian transformation. This has already been done in a certain form by Jacobi, who placed alongside of the modular equation (6), properly so called, a series of other equations of the $(n+1)^{\text{th}}$ degree which can replace it. It is no part of my plan to communicate a rational and comprehensive theory of the infinite number of equations which thus come under consideration.* We must confine our thoughts to an especially important result which Jacobi had established as early as 1829 in his "Notices sur les fonctions elliptiques."† Jacobi there considers, instead of the modular equation, the so-called *multiplier-equation*, together with other equations equivalent to it, and finds that *their* $(n+1)$ *roots are composed in a simple manner of* $\frac{n+1}{2}$ *elements, with the help of merely numerical irrationalities.* Namely, if we denote these elements by $A_0, A_1, \ldots A_{\frac{n-1}{2}}$, and, further, for the roots z of the equation under consideration, apply the indices employed by Galois, we have, with appropriate determination of the square root occurring on the left-hand side:

* *Cf.* for this, so far as modular equations proper are concerned, my developments: "Zur Theorie der elliptischen Modulfunctionen," in Bd. xvii of Mathematische Annalen (1879).

† Crelle's Journal, Bd. iii, p. 308; or Werke, t. i, p. 261.

$$(13) \qquad \begin{cases} \sqrt{z_\infty} = \sqrt{(-1)^{\frac{n-1}{2}} \cdot n} \cdot \mathsf{A}_0, \\ \sqrt{z_\nu} = \mathsf{A}_0 + \epsilon_\nu \mathsf{A}_1 + \epsilon^{4\nu} \mathsf{A}_2 + \ldots \epsilon^{\left(\frac{n-1}{2}\right)^2 \nu} \cdot \mathsf{A}_{\frac{n-1}{2}} \end{cases}$$

for $\nu = 0, 1, \ldots (n-1)$ and $\epsilon = e^{\frac{2i\pi}{n}}$, so that, therefore, the following relations hold good between the $\sqrt{z}$'s:

$$(14) \qquad \begin{cases} \sum_\nu \sqrt{z_\nu} = \sqrt{(-1)^{\frac{n-1}{2}} \cdot n} \cdot \sqrt{z_\infty}, \\ \sum_\nu \epsilon^{-N \cdot \nu} \sqrt{z_\nu} = 0, \end{cases}$$

where N is to denote any one of the $\frac{n-1}{2}$ non-residues for modulus n.

Jacobi has himself emphasised the special significance of his result by adding to his short communication: "C'est un théorème des plus importants dans la théorie algébrique de la transformation et de la division des fonctions elliptiques." Our further report will show how true this remark has proved. In the hands of *Kronecker* and *Brioschi,* the formulæ (13) (14) have attained a general importance for algebra, inasmuch as the savants just mentioned determined to consider Jacobi's equations of the $(n+1)^{\text{th}}$ degree, *i.e.*, therefore equations whose $(n+1)$ roots satisfy the established relations, independently of their connection with the theory of elliptic functions.* But in particular, on the existence of the Jacobian equations of the sixth degree (which correspond to $n=5$), rests Kronecker's theory of equations of the fifth degree, as we shall soon have to show in detail.

§ 4. On Hermite's Work of 1858.

We have now all the preliminary conditions for understanding *Hermite's* first work in this connection, the oft-mentioned

* I follow throughout the notation and nomenclature of Signor Brioschi, as I did in my earlier publications. Herr Kronecker differs particularly in writing $z=f^2$, and thus obtaining equations of the $(2n+2)^{\text{th}}$ degree, whereupon linear identities corresponding to the formulæ (14) exist between the magnitudes f. I do not see that this possesses many advantages.

memoir of April 1858.* Hermite had even earlier been concerned (as also had *Betti*) with the proof of Galois' data concerning the group of the modular equation. But the object was, so far as the case $n=5$ was concerned, to actually establish, in the simplest form, that resolvent of the fifth degree which the modular equation (7) ought to possess. This is what Hermite now attained to, when he put:

$$(15) \qquad y=(v_\infty - v_0)\ (v_1 - v_4)\ (v_2 - v_3),$$

and found the following corresponding equation of the fifth degree:

$$(16) \quad y^5 - 2^4 . 5^3 . u^4 (1-u^8)^2 . y - 2^6 \sqrt{5^5} . u^3 (1-u^8)^2 (1+u^8) = 0.\dagger$$

We have here exactly the *Bring* form with which we became acquainted above, and, in fact, it is easy to identify any Bring equation with (16) by a suitable choice of u. It is sufficient to return to the simplified form which we communicated in (5):

$$t^5 - t - A = 0.$$

We reduce (16) to this form on taking:

$$(17) \qquad y = 2\sqrt[4]{5^3} . u . \sqrt{1-u^8} . t,$$

the coefficient A will then be equal to the following expression:

$$(18) \qquad \frac{2}{\sqrt[4]{5^5}} \cdot \frac{1+u^8}{u^2 (1-u^8)^{\frac{3}{2}}} = A,$$

and here we determine u from A the more easily in that we have to do with a *reciprocal* equation with regard to u. *Hence the solution of any Bring equation is furnished by the formulæ of Hermite, and with it indirectly the solution of the general equation of the fifth degree by means of elliptic functions.*

Hermite's work has, as follows from this short account, no kind of relation to the *algebraical* theory of equations of the fifth degree. Rather it moves throughout in the field of elliptic modular functions, and, moreover, the series of further researches which Hermite has published on the theory of modular

* Comptes Rendus, t. 46: "Sur la résolution de l'équation du cinquième degré."

† For the proof *cf.*, say, *Briot-Bouquet*, "Théorie des fonctions elliptiques" (Paris, 1875), p. 654, &c.

functions took its origin in these. This is the reason why Hermite's solution of the equation of the fifth degree only comes cursorily under consideration in our following exposition; for the use of elliptic functions appears altogether secondary to the conception which we shall henceforward maintain. This would, of course, be at once changed if we wanted to take into account in detail the general ideas which we formulated in the concluding paragraph of the preceding Part, a course which must be deferred to future expositions.

Together with Hermite's first work we advisedly mention two communications of *Brioschi* and *Joubert*, who both compute the resolvents of the fifth degree for the multiplier-equation of the sixth degree (a special Jacobian equation, therefore, of the sixth degree), and hence likewise obtain the equation (16).* Kronecker had also, as he informs Hermite, dealt originally with resolvent construction of this kind.†

§ 5. The Jacobian Equations of the Sixth Degree.

Continuing our account, let us now first turn our thoughts to the researches which *Brioschi* and *Kronecker* have made with regard to Jacobian equations of the sixth degree.‡ Let us first remark the following facts. Whenever two investigators have worked at the same subject simultaneously and in relation to one another, it is difficult to distinguish what was discovered by the one, what by the other. The chronological method, which refers to the dates of the individual publications, is certainly not always appropriate; but it is, after all, the only one which can be handled with any certainty. In this sense we shall now proceed on the basis of this method. I begin with recounting the works which *Signor Brioschi* has published in the first volume of the Annali di Matematica, Serie I (1858).

* Brioschi: "Sulla risoluzione delle equazioni di quinto grado" (Annali di Matematica, Ser. I, t. i, June 1858), *Joubert* in a communication from Hermite in vol. 46 of the Comptes Rendus ("Sur la résolution de l'équation du quatrième degré," April 1858). See also Joubert: "Note sur la résolution de l'équation du cinquième degré," in the Comptes Rendus, t. 48 (1859).

† Letter to Hermite, June 1858. See Comptes Rendus, t. 46.

‡ Compare the exposition of this relation by *Hermite* in his memoir, already mentioned: "Sur l'équation du cinquième degré," Comptes Rendus, particularly t. 62 (1866), pp. 245–247.

After Signor Brioschi had first proved * (l. c.) the data of Jacobi, he concerned himself with the actual establishment of the general Jacobian equation of the sixth degree. His result is as follows.† Let A_0, A_1, A_2 be three magnitudes which occur in (13) corresponding to $n=5$; further, let:

$$(19)\quad \begin{cases} A = A_0^2 + A_1A_2, \\ B = 8A_0^4A_1A_2 - 2A_0^2A_1^2A_2^2 + A_1^3A_2^3 - A_0(A_1^5 + A_2^5), \\ C = 320A_0^6A_1^2A_2^2 - 160A_0^4A_1^3A_2^3 + 20A_0^2A_1^4A_2^4 + 6A_1^5A_2^5 \\ \quad - 4A_0(A_1^5 + A_2^5)(32A_0^4 - 20A_0^2A_1A_2 + 5A_1^2A_2^2) \\ \quad + A_1^{10} + A_2^{10}. \end{cases}$$

Then the general Jacobian equation of the sixth degree will be the following:

$$(20)\quad (z-A)^6 - 4A(z-A)^5 + 10B(z-A)^3 - C(z-A) + (5B^2 - AC) = 0.$$

Brioschi further seeks to construct a resolvent of the fifth degree as simple as possible for this equation, and to this end first ‡ puts (following Hermite's example):

$$(21)\qquad y = (z_\infty - z_0)(z_1 - z_4)(z_2 - z_3),$$

but then remarks, in connection with a letter of Hermite's, that the square root of this expression is already rational in the A's, and gives rise to an equation of the fifth degree.§ Let x be this square root; then Brioschi finds for the five values of which x is susceptible the following formulæ:

$$(22)\qquad \begin{aligned} x_\nu = & -\epsilon^\nu A_1(4A_0^2 - A_1A_2) + \epsilon^{2\nu}(2A_0A_1^2 - A_2^3) \\ & + \epsilon^{3\nu}(-2A_0A_2^2 + A_1^3) + \epsilon^{4\nu}A_2(4A_0^2 - A_1A_2), \end{aligned}$$

while for the corresponding equation of the fifth degree he finds this:

$$(23)\qquad x^5 + 10Bx^3 + 5(9B^2 - AC)x - \sqrt[4]{\frac{\Pi}{5^5}} = 0,$$

where Π is the discriminant of the Jacobian equation (20).‖

The *multiplier-equation of the sixth degree* for elliptic functions (to which Jacobi's remark first related) is of course contained in (20) as a special case. Brioschi finds that it is

* P. 175, l. c. (May 1858).

† P. 256, l. c. (June 1858).

‡ Loc. cit.

§ P. 326, l. c. (Sept. 1858).

‖ I have here, in opposition to the original formula of Brioschi, given the numerical coefficients, as Joubert had done later on ("Sur l'équation du sixième degré," Comptes Rendus, t. 64, 1867).

essentially characterised by the condition $B=0$, whereupon (23) becomes a Bring equation. *To Herr Kronecker is due the credit of first directing attention to the case $A=0$, and also of effecting its solution by means of elliptic functions.* We need not communicate here in detail his primary formulæ as he noted* them in his letter to Hermite, and as Brioschi then proved them in the memoir (to be presently described more in detail) in the first volume of the Atti of the Istituto Lombardo.† For they are considerably simplified if, instead of the modulus k (which Herr Kronecker used), the rational invariants of the elliptic integral g_2, g_3, Δ are introduced, and we have already become acquainted (I, 5, § 8) with the formulæ of solution in question in this simplified form. *In fact, the Jacobian equation of the sixth degree with $A=0$ is none other than that simplest resolvent of the sixth degree which we have established in* I, 4, § 15, *in the case of the icosahedron.* We have only to put:

$$\text{(24)} \qquad \mathsf{A}_0 = z_1 z_2,\ \mathsf{A}_1 = z_1^2,\ \mathsf{A}_2 = -z_2^2,$$

and correspondingly:

$$\text{(25)} \qquad B = -f,\ C = -H.$$

At the same time, for $A=0$, the resolvent of the fifth degree (23) is transformed into the following:

$$\text{(26)} \qquad x^5 + 10Bx^3 + 45B^2x - \sqrt[4]{\frac{\Pi}{5^5}} = 0,$$

which agrees with formula (27) of I, 4, § 11. I mention these relations only cursorily, to return to them later more in detail. It remains to consider one final direction of investigation with regard to the Jacobian equations of the sixth (or indeed of any) degree, that which *Herr Kronecker* first took in hand ‡ in his algebraical communications from the year 1861 onwards, and which was then followed up further by Signor Brioschi in particular in the first volume of the second series of the Annali di Matematica (1867).§ The object is to construct from one

* Comptes Rendus, t. 46, June 1858.

† "Sul metodo di Kronecker per la risoluzione delle equazioni di quinto grado" (Nov. 1858).

‡ Monatsberichte der Berliner Akademie.

§ "La soluzione più generale delle equazioni del 5. grado." See also "Sopra alcune nuove relazioni modulari," in the Atti della R. Accademia di Napoli of 1866.

Jacobian equation a new one by a Tschirnhausian transformation. Herr Kronecker remarks that this is possible in two ways, inasmuch as the roots Z_∞, Z_ν of the transformed equation (which correspond to the z_∞, z_ν of the original equation) either just satisfy the formulæ (13), (14) (where ϵ can be replaced by ϵ^R at pleasure, understanding by R a quadratic residue of n; this only signifies a change in the order of the roots); *or they satisfy those others which proceed from* (13), (14), *on replacing* ϵ *by* ϵ^N, *where N is to denote an arbitrary non-residue to the modulus n.* Let n be, as we will now assume, equal to 5; then we can in the first case put $\sqrt{Z}$ equal, for example, to $\frac{\delta\sqrt{z}}{\delta A}$ or equal to $\frac{\delta\sqrt{z}}{\delta B}$; the most general expression for $\sqrt{Z}$ here coming under consideration arises on combining $\sqrt{z}$ and the two magnitudes mentioned multiplied by arbitrary constant factors:

$$(27) \qquad \sqrt{Z}=\lambda\,.\,\sqrt{z}+\mu\,.\,\frac{\delta\sqrt{z}}{\delta A}+\nu\,.\,\frac{\delta\sqrt{z}}{\delta B}.$$

We solve the second case on first constructing for it a particular example, which is furnished, say, by:

$$(28) \qquad Z=\frac{1}{z-A}+\frac{C}{5B^2-AC};$$

afterwards we treat the Jacobian equation corresponding to this example exactly according to formula (27). We shall return later on more in detail to the principle of these transformations. Meanwhile let us find room for the following remark. If we calculate for the $\sqrt{Z}$ of formula (27) the expression A, this will be an integral homogeneous function of the second degree of the λ, μ, ν. We can make this zero by, for instance, putting $\nu=0$ and determining $\lambda:\mu$ by means of the resulting quadratic equation. *We can therefore by mere extraction of a square root transform the general Jacobian equation of the sixth degree into one with $A=0$.*

Signor Brioschi has since collected * his researches here indicated, as also the further ones to be described presently,

* "Ueber die Auflösung der Gleichungen fünften Grades" (1878).

which relate specially to the theory of equations of the fifth degree, in Bd. xiii of the Mathematische Annalen; and they are all the more welcome because his original publications, widely scattered as they were, could have been only with difficulty accessible to many mathematicians. *Herr Kronecker* has also since returned to the theory of the general Jacobian equations,* but the questions there treated by him lie beyond the limits which are prescribed for our present exposition.

§ 6. Kronecker's Method for the Solution of Equations of the Fifth Degree.

Having premised the theory of the Jacobian equations of the sixth degree, we can with ease describe the nature of that method of solution which *Herr Kronecker* has developed in his oft-cited letters to Hermite (Comptes Rendus, t. 46, June 1858) for the general equation of the fifth degree. The Jacobian equations of the sixth degree are very intimately bound up with the theory of elliptic functions, but they also represent, as we have already remarked (and this in virtue of formulæ (13), (14)), a remarkably simple type of algebraical irrationalities *per se.* Herr Kronecker's particular discovery is this: *that from the general equation of the fifth degree after adjunction of the square root of the discriminant, rational resolvents of the sixth degree can be established which are Jacobian equations.* To this is appended the further remark, which we led up to just now: *that we can transform the Jacobian equation in question by the help of only one additional square root into one with* $A=0$, *therefore into a normal form with only one essential parameter,*† *which admits of solution by elliptic functions.*

In Herr Kronecker's original communication the two points here separated are, however, not clearly distinguished. Herr Kronecker limits himself to communicating the following rational function of the five roots of an equation of the fifth degree:

* Monatsberichte der Berliner Akademie of 1879: "Zur Theorie der algebraischen Gleichungen."

† Here again we reduce it to only one parameter by putting $z=\rho t$ and determining ρ suitably.

$$f(\nu, x_0, x_1, x_2, x_3, x_4)$$

(29) $$= \sum_{m=0}^{m=4} \sum_{n=0}^{n=4} \sin \frac{2n\pi}{5}(x_m x^2_{m+n} x^2_{m+2n} + \nu x^3_m x_{m+n} x_{m+2n}),$$

in which he supposes ν so determined that $\Sigma f^2 = 0$, and then remarks that the several f's, which arise from (29) by even permutations of the x's, satisfy an equation of the twelfth degree of the following form:

(30) $$f^{12} - 10\phi \,.\, f^6 + 5\psi^2 = \psi \,.\, f^2,$$

which will admit of solution with the help of elliptic functions. Here (30), provided we put $f^2 = z$, is the Jacobian equation with $A = 0$, and the vanishing of A corresponds to the vanishing of Σf^2.

We are indebted to Signor Brioschi for having made the deeper meaning of Kronecker's method accessible to the mathematical public in a lucid and at the same time a more general form, and this in the memoir mentioned just now: "*Sul metodo di Kronecker*," &c., in the first volume of the Atti of the Istituto Lombardo (Nov. 1858). We do not here recur to the contributions which Brioschi has there made to the general theory of the Jacobian equations of the sixth degree. What here interests us is that *he establishes a general rule of construction for the roots z, of which a special case occurs in formula* (29). Let:

(31) $$v(x_0, x_1, x_2, x_3, x_4)$$

be a rational function of the five x's which remains unaltered for the cyclic permutation:

$$(x_0, x_1, x_2, x_3, x_4);$$

further, let:

(32) $$v' = v(x_0, x_4, x_3, x_2, x_1),$$

Brioschi then puts:

(33) $$v - v' = u_\infty,$$

and derives from this function five new functions u_0, u_1, u_2, u_3, u_4, by first subjecting the x's to the substitution:

$$x_0' = x_0,\ x_1' = x_3,\ x_2' = x_1,\ x_3' = x_4,\ x_4' = x_2,$$

and then bringing into application the cyclic permutation already mentioned. *Then the following expressions are in general found to be the roots of a Jacobian equation of the sixth degree, which remains unaltered for all even permutations of the x's, and hence possesses as coefficients rational functions of the coefficients of the equation of the fifth degree and of the square root of its discriminant:*

$$(34)\qquad \begin{cases} z_\infty = (u_\infty\sqrt{5} + u_0 + u_1 + u_2 + u_3 + u_4)^2, \\ z_0 = (u_\infty + u_0\sqrt{5} - u_1 + u_2 + u_3 - u_4)^2, \\ z_1 = (u_\infty - u_0 + u_1\sqrt{5} - u_2 + u_3 + u_4)^2, \\ z_2 = (u_\infty + u_0 - u_1 + u_2\sqrt{5} - u_3 + u_4)^2, \\ z_3 = (u_\infty + u_0 + u_1 - u_2 + u_3\sqrt{5} + u_4)^2, \\ z_4 = (u_\infty - u_0 + u_1 + u_2 - u_3 + u_4\sqrt{5})^2. \end{cases}$$

These formulæ become still more concise if we note the elements A_0, A_1, A_2, of which the $\sqrt{z}$'s, in accordance with (13), are composed. The comparison gives simply:

$$(35)\qquad \begin{cases} A_0\sqrt{5} = u_\infty\sqrt{5} + u_0 + u_1 + u_2 + u_3 + u_4, \\ \frac{1}{2}A_1\sqrt{5} = u_0 + \epsilon^4 u_1 + \epsilon^3 u_2 + \epsilon^2 u_3 + \epsilon u_4, \\ \frac{1}{2}A_2\sqrt{5} = u_0 + \epsilon u_1 + \epsilon^2 u_2 + \epsilon^3 u_3 + \epsilon^4 u_4, \end{cases}$$

where $\epsilon = e^{\frac{2i\pi}{5}}$, $\sqrt{5} = \epsilon + \epsilon^4 - \epsilon^2 - \epsilon^3$. The formulæ (29) are, as we have already pointed out, included in (34) as a special case. Herr Kronecker has here from the first endowed the functions v or u which he used with a parameter ν occurring linearly, in order to be able to satisfy the additional condition $A=0$. Signor Brioschi gives, for another case connected with the invariants of the binary form of the fifth degree, the full calculation of the final equation of the sixth degree.

We have just become acquainted in (23) with Brioschi's simple resolvent of the fifth degree for the Jacobian equation of the sixth degree. Now considering the Jacobian equation of the sixth degree in its turn as a resolvent of the general equation of the fifth degree, we recognise the possibility of *transforming the general equation of the fifth degree by means of a Tschirnhausian transformation, whose coefficients are rational after adjunction of the square root of the discriminant of the* *proposed equation, into an equation* (23), i.e., *an equation in which the fourth and the second power of the unknown are wanting.** In particular, if we annex, besides, Kronecker's auxiliary equation for ν, we can make $A=0$ in this equation, and thus obtain the form (26), which, like the Bring form, only contains one essential parameter. *Hermite*, and after him *Brioschi* again, have dealt in detail with the problem of constructing the Tschirnhausian transformation in question in an explicit form. We should have to go into these works more minutely, if it were not (as we have said) that they are essentially controlled by the requirement: to bring into play the invariants of the binary form of the fifth order. Let us therefore only briefly refer here, first, to the elegant communication which Hermite makes to Borchardt in Bd. 59 of the Journal für Mathematik (1861); then to his oft-mentioned exhaustive memoir, *Sur l'équation du cinquième degré*, of which the second half (Comptes Rendus, t. 62 (1866), pp. 715, 919, 959, 1054, 1161) is devoted to the exact accomplishment of all the calculations which appear necessary in Kronecker's method; and finally to a series of remarks which Signor Brioschi has then appended to the developments of Hermite. (Comptes Rendus, t. 63 [1866, 2], t. 73 [1871, 2], t. 80 [1875, 1]).†

§ 7. On Kronecker's Work of 1861.

Though Herr Kronecker in his first communication to Hermite had only cursorily and by an example, so to say, demonstrated his method of solution of equations of the fifth degree, he has since (1861) gone into‡ the nature and principles of it more thoroughly. Our account of it must be the more complete in this place because the reflexions in question in many respects lie at the root of our own developments in

* The mode of expression in the text premises what we shall presently remark concerning the irrationality of $\sqrt[4]{\frac{11}{5^5}}$.

† *Cf.* also M. Roberts in the first volume of the 2nd series of the Annali di Matematica (1867): "Note sur les équations du cinquième degré."

‡ Namely, in the already-mentioned communication in the Berliner Monatsberichten, of which that part which relates to equations of the fifth degree was reprinted in Bd. 59 of Borchardt's Journal.

the sequel, and on the other hand Herr Kronecker has given a peculiarly scanty exposition of them, omitting all proofs.

First, Herr Kronecker expressly distinguishes between the transcendental and the algebraical part of the solution. The latter, the more particularly important, consists in the assemblage of all those algebraical operations which are necessary in order to replace the general equation of the fifth degree by a *normal form*, the simplest that can be chosen: how we elect to calculate the roots of this latter given case by convergent infinite processes, or by empirical tables, or what not, is a question *per se* which is not further touched upon. Hence the Jacobian equations of the sixth degree for Herr Kronecker now only come under consideration in virtue of their algebraical peculiarities, not in virtue of their connection with elliptic functions.

Secondly, Herr Kronecker remarks that we must draw an essential distinction between the irrationalities which are introduced for the purposes of the reduction of algebraical equations. The irrationalities of the first—we might call them the *natural* ones—are those which depend rationally on the roots x which are to be determined, the same, therefore, as we have described in the fourth chapter of the preceding Part as roots of "rational" resolvents. Alongside of these appear the others, which we might call *accessory*, because they are irrational functions of the x's. Such accessory irrationalities need not be more complicated than the natural ones, *e.g.*, they may involve the square root of a coefficient of the proposed equation. This is the case with the expressions (29) which we just considered; these in themselves denote natural irrationalities, which, however, become accessory if the ν is determined in the way explained with the help of a quadratic equation.

In accordance with this distinction, Herr Kronecker further asks to what point we can go in the solution of equations of the fifth degree when we impose the restriction of only wishing to employ *natural* irrationalities. The Jacobian equation of the sixth degree contains primarily *three* parameters, to wit, the three magnitudes which we have denoted by A, B, C. Herr Kronecker remarks that by appropriate modification of his method, without leaving the circle of the natural irrationalities, we can replace these parameters by two only, a and b. *On the other hand*, he asserts, *it is impossible, without accepting the aid of accessory irrationalities, to construct from the general equation of the fifth degree a Jacobian equation with only one parameter, or any resolvent at all with only one parameter.*

The first of these two points may be immediately established by calculation. We will show, viz., in the fourth chapter following, that, alongside of the expressions of the second, sixth, and tenth degrees in A_0, A_1, A_2, which we called A, B, C, yet another expression of the fifteenth degree, D, is rationally known, whose square is an integral function of the A, B, and C. We have already encountered this D as the fourth root of the discriminant (divided by 5^5) of the Jacobian equation, in the constant term of (23). Let us now, in the resolvent of the equation of the fifth degree, replace the expressions A_0, A_1, A_2 (35), by $\frac{\mathsf{A}_0 \cdot A^7}{D}$, $\frac{\mathsf{A}_1 \cdot A^7}{D}$, $\frac{\mathsf{A}_2 \cdot A^7}{D}$, *i.e.*, by functions of null dimensions proportional to them. Thus, in the place of A, B, C, D, appear respectively $\frac{A^{15}}{D^2}$, $\frac{A^{42} \cdot B}{D^6}$, $\frac{A^{70} \cdot C}{D^{10}}$, $\frac{A^{105}}{D^{14}}$. Here we can substitute for D throughout the integral function of A, B, C which is equal to it. Then the new A, B, C, D depend, in fact, on only two parameters, viz., the quotients of null dimensions:

$$(36) \qquad a = \frac{B}{A}, \quad b = \frac{C}{A^5},$$

whereupon the required proof is achieved.

The proof of the second assertion is essentially more difficult, and we must defer it till the conclusion of our main exposition. It there appears as a consequence of properties of the icosahedral substitutions which we have before brought into prominence, and flows so naturally from them that the real basis of the theorem in question seems to be disclosed by means of them.

I come to the conclusion of Kronecker's work. Herr Kronecker calls attention to the fact that, in the case of those algebraic equations which admit of solution by the extraction of roots, and indeed on the ground of the original developments of *Abel*, the accessory irrationalities can be dispensed with altogether. He then postulates the same for the *solution of higher equations: he only wants to see their reduction brought* *in each case so far as the use of the natural irrationalities carries it.* This is, therefore, the last step of the original method of Kronecker, as we have just become acquainted with it: to push the reduction back to an equation with $A=0$. Or rather, the theory has to confine itself to placing the equations of the fifth degree (in the way just suggested) in connection with Jacobian equations which contain two parameters; to investigate the different kinds of reduction here possible; and, finally, to see how, conversely, the roots of the equation of the fifth degree are now represented in terms of the roots of the said Jacobian equation of the sixth degree.*

As regards our own exposition, I should like, in the sequel, *not* to retain the requirement here detailed. True, we shall have to investigate—and this shall be done in the fullest manner—how far we can get with the use of natural irrationalities only. But, over and above this, the question arises, what is the state of affairs with regard to the accessory irrationalities which aid us in the further reduction; what are the simplest results which we can attain to with their help? The analogy with those equations which are solvable by extraction of roots does not seem to me to have much force. If, for the latter, the use of accessory irrationalities is superfluous, we may perceive in the necessary occurrence of these irrationalities, in the case of higher equations, a characteristic feature of these latter, and should rather proceed, in the case of equations of the fifth degree, as the lowest case of higher equations, to fathom the nature and significance of the necessary accessory irrationalities. We shall the less be able to neglect these investigations, because the treatment of the natural irrationalities is, as we shall see, in a certain sense furnished by them.

* Here I should like to direct attention afresh to the concluding paragraph of I, 5. If the illustrations there given are accurate, we can consider the use of elliptic functions as an introduction of accessory irrationalities of infinitely high order. If we, therefore, wish to retain Kronecker's postulate, it is no use trying to proceed to solve these equations with two parameters which we have obtained, for these equations form a point beyond which further advance is impossible.

§ 8. Object of our further Developments.

At this point we break off our historical account, inasmuch as it seems to the purpose to interweave the description of the works still to be mentioned with the progressive exposition of the following chapter.* The object of this exposition is, as we have repeatedly pointed out, to place the solution of equations of the fifth degree in connection with the theory of the icosahedron in a manner as simple and comprehensive as possible. That such a connection is possible follows in several ways from the exposition which we have so far given: for the Jacobian equation with $A=0$ is, as we saw, a resolvent of the icosahedral equation; and we can even conceive the Bring form as such, if we suppose in I, 4, § 12, the ratio $m:n$ so determined that the term involving Y^2 in the canonical resolvent vanishes.

However, it is not our intention to introduce the icosahedron in such an indirect manner. We desire rather to expound the theory of equations of the fifth degree connectedly, and in such a manner from the outset that the significance of the icosahedron will be recognised as necessary and fundamental. I here employ freely constructions in the sense of *projective geometry*, as I have already repeatedly noted. No doubt we can throughout replace these by purely algebraical reflexions. Nevertheless I believe that they are essentially useful, and am of opinion that they must be also of importance in a similar form in higher problems of the theory of equations.

* These are first the different essays which have been published by Herr Gordan under the title, "Ueber die Auflösung der Gleichungen fünften Grades," and by myself as "Weitere Untersuchungen über das Ikosaeder." The first are found respectively in the Erlanger Berichten of July 1877, in the official report of the Naturforscherversammlung at München (Sept. 1877), and in Bd. xiii of the Math. Annalen (1878); the latter in the Erlanger Berichten of Nov. 1876, January and July 1887, and, finally, in Bd. xii of the Annalen (1877). See also a communication from Brioschi to the R. Accademia dei Lincei of Dec. 1876 (Transunti), and another to the Istituto Lombardo of April (1877) (Rendiconti (2), X). To this add, further, Kiepert's "Auflösung der Gleichungen fünften Grades" in the Göttinger Nachrichten of July 1878, completed in Borchardt's Journal, t. 87, Aug. 1878; also my own works: "Ueber die Transformation der elliptischen Functionen und die Auflösung der Gleichungen 5. Grades" (Bd. 14 of the Annalen, May 1878), and "Ueber die Auflösung gewisser Gleichungen von 7. und 8. Grade" (Bd. 15 of the Annalen, March 1879).

The details of our following exposition are distributed in four chapters.

Our first object is *to bring the main idea of the theory of equations into a geometrical form.* Here I adopt a mode of exposition which I gave in 1871 in the fourth volume of the Mathematische Annalen,* and develop in particular, by pursuing this further, the geometrical conception of the Tschirnhausian transformation and of resolvent construction. With a view to what follows I append thereto a short excursus on the elements of line geometry and the corresponding properties of the surface of the second degree.

The following third chapter is devoted to the special theory of the *canonical equations of the fifth degree, i.e.*, those equations which contain neither the fourth nor the third power of the unknown. On the basis of the theorem that surfaces of the second degree possess two systems of rectilinear generators, there arises for the said equations a peculiarly simple connection with the icosahedron, whence our earlier developments concerning the *canonical* resolvent of the icosahedral equation (I, 4, § 12) lead to explicit formulæ for the roots of the proposed equation. By this means we obtain in particular, as I develop cursorily, the means of putting the Bring transformation into a definite shape and understanding its real essence.

Our fourth chapter then explains the position of the icosahedron in the theory of *the general Jacobian equations of the sixth degree.* It is shown that the latter, in the sense of I, 5, § 4, represents *a ternary form-problem*, and indeed such as arises from the binary icosahedral problem hitherto considered, by a certain simple process of translation. In the same way, all the manifold results, which we have attained in the theory of the Jacobian equation of the sixth degree, present themselves as it were spontaneously and in part in an improved form. In particular, I shall expound how we accomplish the solution of the general Jacobian equation, after adjunction of an accessory square root, most effectively by the help of the icosahedral equation.

Two ways are now open, as we conclude in the fifth chapter, for solving the general equation of the fifth degree by means

* "Ueber eine geometrische Interpretation der Resolventen algebraischer Gleichungen."

of the icosahedral equation, inasmuch as we are at liberty, viz., either to transform the given equation by a Tschirnhausian transformation into a canonical resolvent of the fifth degree, or by construction of resolvents to place it in connection with the ternary form-problem just described. The one gives, if we like to say so, a simplification of the method of Bring, the other a modification of that of Kronecker. But, at the same time, we recognise that the operations which are used in the two methods differ not in their nature, but only in regard to their order. We thus have the means of comprehending the whole of the older works described in the preceding paragraphs from one point of view. And here we also succeed in providing that indirect theorem, established by Herr Kronecker, of which we just now gave an account, and which can be conceived as a fundamental conclusion not only of the problem of solution in its abstract form, but also specially of our own considerations.

It is, perhaps, particularly interesting that, in virtue of our exposition, the theory of equations of the fifth degree is again brought near to that of equations of the third and fourth degree. We have paid regard to this, wherever it seemed useful, in brief footnotes.

CHAPTER II

INTRODUCTION OF GEOMETRICAL MATERIAL

§ 1. Foundation of the Geometrical Interpretation.

The geometrical interpretation of equations of the fifth degree, with which we shall work in the following pages, rests on the simple idea of using the roots x_0, x_1, x_2, x_3, x_4 of the equation as homogeneous point-coördinates (where, of course, only the ratios of the x's are actually interpreted). Were we not to add hereto a further limitation, we should have to start from a space of four dimensions. But this would be doubly inconvenient: we should have to forego the pregnant terminology which is at our disposal for space of three dimensions, and should be unable to assume results in a specific form. We will therefore introduce a limitation which will be effected in every case by an easy auxiliary transformation, viz., *by laying down the condition that in what follows we are always to take:*

(1) $$\Sigma x = 0,$$

and that, therefore, we shall only consider equations of the fifth degree of the form:

(2) $$x^5 + ax^3 + bx^2 + cx + d = 0$$

(in which the term in x^4 is wanting). We can then, and in fact immediately in virtue of (1), denote the ratios of the x's as point-coördinates of ordinary space, its so-called *pentahedral co-ordinates.* Such pentahedral co-ordinates are only formally different from the ordinary tetrahedral co-ordinates of projective geometry; we might define them directly in this way, that we consider four of them as tetrahedral co-ordinates, and introduce the fifth in virtue of (1) as a linear combination of the rest; only the symmetry on which we lay the greatest weight in the sequel is then lost.*

* The introduction of superfluous co-ordinates, which are then connected by a corresponding number of linear identities, is otherwise useful in geo-

The geometrical interpretation here described derives its primary importance from our considering the different *arrangements* which we can impart to the roots x. To one and the same equation of the fifth degree correspond in this sense at the outset 120 points of space, in general distinct, which are only known in the aggregate; the solution of the equation will then consist in supplying the means of distinguishing the individual points from the 120 introduced in this form.

The points spoken of here are, of course, not geometrically independent. An arbitrary permutation of the pentahedral co-ordinates, *e.g.*, that which replaces x_k by x_i, can be denoted geometrically as a *transformation of the whole space*, viz., as that collineation thereof which corresponds to the formula:

(3) $$x_i' = x_k.$$

The 120 collineations which correspond in this sense to the 120 permutations of the x's are clearly defined geometrically by the fact *that they all transform into itself the pentahedron which determines the co-ordinates.* The geometrical connection of the 120 associated points is just this, that they all proceed from one of themselves by means of the said collineations.

I have here restricted the development of these fundamental ideas to the equations of the fifth degree. This restriction, however, is quite unimportant; a perfectly analogous kind of geometrical meaning is possible for equations of the nth degree, provided we start from projective space of $(n-2)$ dimensions; thus, for equations of the fourth degree, from the plane; for equations of the third degree, from the straight line. We can here indeed take account of the Galois-affect of the equations by considering, instead of the possible permutations of the n-roots and the collineations corresponding to them, only a sub-group thereof. It is unnecessary in what follows to treat the matter under such general conditions. Howbeit, I might just point out here the perfectly similar geometrical meaning which we shall use in the next chapter but one, in our investigation of the form-problem there discussed.

metry. *Cf.*, *e.g.*, Paul Serret's "Géométrie de direction" (Paris, 1869). The system of pentahedral co-ordinates in particular was, I believe, first used by Hamilton in his researches on the geometrical net of Möbius, which can be derived from five points in space. See Hamilton's "Elements of Quaternions" (Dublin, 1866), pp. 57–77.

§ 2. Classification of the Curves and Surfaces.

Let us observe, moreover, that we can classify the curves and surfaces of our space (or in general the geometrical figures existing therein) according to their behaviour with regard to the 120 collineations (3). In general, an irreducible curve or surface is not transformed into itself by any of the 120 operations; it appears then as one of the 120 associated figures, of which each possesses the same properties both in itself and in relation to the co-ordinate pentahedron. But it can also be transformed into itself by the n-transformations of a determinate sub-group g contained in the aggregate 120 transformations. Then the number of the co-ordinated figures is only $\frac{120}{n}$; each one remains unaltered by the n-transformations of a sub-group which is associated with the group g within the main group. Evidently the same distinctions occur here which we found in the fourth chapter of the first Part in treating of the theory of resolvent construction.

We will introduce a definite terminology in connection with this. If a figure is transformed into itself by all the 120 collineations, we call it *regular; half-regular*, on the other hand, if this is only the case with regard to the 60 collineations which correspond to the even permutations of the x's, and which we may denote shortly as the even collineations. In all other cases we shall speak of *irregular* figures. The half-regular figures group themselves together naturally in pairs, for the group of the 60 even collineations in self-conjugate within the main group; therefore if one figure is transformed into itself by the 60 even collineations, so will also the other be which proceeds from it by an arbitrary uneven collineation.

The classification here described will be of importance for the purposes of the theory of equations, inasmuch as we now consider equations which contain *parameters*. We will only count these parameters as they affect the ratios $x_0 : x_1 : x_2 : x_3 : x_4$. Then, if we have an equation with *one* parameter, the 120 corresponding points x trace out, by the variation of the parameter, a curve in space which will be transformed into itself by the 120 collineations, and which we shall call the image of the equation. Similarly, we obtain, as the image of the equation, a surface when the number of essential parameters is two; the surface is also transformed into itself by the 120 collineations. The question whether this curve or surface is reducible or not is evidently intimately connected with the *group* of the given equation of the fifth degree. To fix the ideas, I shall assume that our parameters enter rationally in the coefficients of the equation. At the same time, we will lay no stress on mere numerical irrationalities; we shall, therefore, regard arbitrary rational functions of the parameters as rationally known. Then the Galois group of the equation [in conformity to I, 4] is transformed into that which Hermite * has called, by way of definition, the *group of monodromy, i.e.*, the aggregate of those permutations of the roots x which occur when we consider the x's as algebraic functions of the parameters, and then let these, starting from any initial values, so vary in the complex domain that they finally return to their initial values. The point x moves by this process of variation in the same irreducible portion of the geometrical image corresponding to the equation, and assumes therein, by suitable variation of the path, all possible positions. We conclude from this *that the irreducible portion in question is transformed into itself by just so many collineations among the* 120 *which exist on the whole as there are permutations of the x's contained in the group of monodromy.* It will not be difficult to confirm this general proposition in the particular examples which we now enter upon.

§ 3. The Simplest Special Cases of Equations of the Fifth Degree.

With a view to our later developments, we now consider the simplest special cases of equations of the fifth degree, namely, those which proceed from (2) by equating one or more coefficients to zero, in which the other coefficients (in so far as they influence the relations of the roots x) will have to figure as parameters.

* Comptes Rendus, t. xxxii (1851): "Sur les fonctions algébriques;" see, too, C. Jordan, "Traité des substitutions," &c., p. 227, &c.

First, let $a=0$; then we have, by (1):*

(4) $$\Sigma x^2=0,$$

i.e., an equation which represents a surface of the second order. If we eliminate by means of (1) the x_4, and form the discriminant from the left side of the equation then existing:

$$x_0^2+x_1^2+x_2^2+x_3^2+(x_0+x_1+x_2+x_3)^2=0,$$

we arrive at $+5$, a value, therefore, which does not vanish. We conclude from this that our surface of the second degree not only does not split up, but is never a cone. It is this surface—*regular* in the sense agreed upon—which will play the most important part in our further geometrical developments. I shall therefore describe it as the Canonical Surface, consistently with the fact that we have already called an equation which satisfies the relations (1) and (4), a Canonical Equation.

We proceed to the following case: $b=0$. Again making use of (1), we obtain for the corresponding x's:

(5) $$\Sigma x^3=0.$$

We are therefore led to that irreducible surface of the third order which Clebsch has incidentally described as the *Diagonal Surface*,† because it contains the diagonals of the co-ordinate pentahedron, *i.e.*, those fifteen lines which, moving in one of the five pentahedral planes, connect any two opposite angles of the quadrilateral marked out in this plane by the other co-ordinate planes. An equation with $b=0$ is, accordingly, to be described in the following pages as a *diagonal* equation. The general Brioschi resolvent, which we have become acquainted with in § 5 of the preceding chapter [formula (23)], is at the same time the general diagonal surface, a circumstance to which we shall return more in detail.

* We recall in what follows the formulæ of Newton, which connect the coefficients of the equation with the sums of the powers $s_\nu=\Sigma x^\nu$. For our equation (2) these formulæ become:

$$s_1=0,\ s_2+2a=0,\ s_3+3b=0,\ s_4+as_2+4c=0, \text{ \&c., \&c.}$$

† See the essay (which will be again quoted): "Ueber die Anwendung der quadratischen Substitution auf die Gleichungen 5. Grades und die geometrische Theorie des ebenen Fünfseits," in Bd. iv of the Math. Ann. (1871). The diagonal surface has otherwise become of importance in the theory of surfaces of the third degree; consult, *e.g.*, my work, "Ueber Flächen dritter Ordnung," in Bd. vi of the Mathematische Annalen (1873).

We next put $a=0$, $b=0$, at the same time. Then the relations (1), (4), (5) hold simultaneously, while the equation (2) assumes Bring's form. *Bring's equations will be therefore represented by the curve of intersection of the canonical surface and the diagonal surface.* In general, a surface of the second and a surface of the third order intersect in an irreducible curve of the sixth order, and of deficiency 4.* We shall show later on that these properties present themselves unaltered in Bring's curve. Bring's curve is therefore certainly regular, just as the canonical surface and diagonal surface are.

The other cases follow in which at least one of the coefficients c, d vanishes. We will not here treat of these individually in detail, inasmuch as we have not to enter specially into the consideration of them. We would only note here that, in the case $d=0$, the figure in space would break up into irregular components, these being the five planes of the co-ordinate pentahedron which correspond to the case $d=0$.

§ 4. Equations of the Fifth Degree which Appertain to the Icosahedron.

We return now to the consideration of those equations of the fifth degree which we have established in the fourth chapter of the preceding Part as resolvents of the icosahedron equation, and seek to arrange them in accordance with the ideas just developed. They are equations with only one essential parameter Z (on the right side of the icosahedron equation), which are therefore to be denoted by *curves*. These curves split up, as we shall show more precisely, into two regular portions. In fact, the group of monodromy is given in every case by the sixty icosahedron substitutions.

Let us begin now, say, with the so-called resolvent of the u's of I, 4, § 11:

(6)
$$48u^5(1-Z)^2-40u^3(1-Z)+15u-4=0.$$

Calculating the sums of the powers, we find:

$$s_1=0,\ s_2=\frac{5}{3(1-Z)},\ s_3=0,\ s_4=\frac{5}{36(1-Z)^2},\ \&c.,$$

* See, *e.g.*, Salmon-Fiedler's "Analytical Geometry of Space" (3rd edition, Teubner, 1880).

therefore:

$$s_2^2 = 20 s_4. \tag{7}$$

We obtain from this as geometrical image of (6) *a curve of the twelfth order which is the intersection of the diagonal surface with the surface of the fourth order* (7). I say, now, that this curve splits up into two half-regular portions of the sixth order, of which each represents a *rational* curve in space. In fact, the roots u_ν of (6), apart from their arbitrary arrangement, are proportional to the octahedral forms previously introduced:

$$t_\nu(z_1, z_2) = \epsilon^{3\nu} z_1^6 + 2\epsilon^{2\nu} z_1^5 z_2 - 5\epsilon^{\nu} z_1^4 z_2^2 - 5\epsilon^{4\nu} z_1^2 z_2^4 - 2\epsilon^{3\nu} z_1 z_2^5 + \epsilon^{2\nu} z_2^6, \tag{8}$$

where z_1, z_2 are connected with Z by the icosahedral equation:

$$\frac{H^3(z_1, z_2)}{1728 f^5(z_1, z_2)} = Z. \tag{9}$$

If Z is an arbitrary variable, so is $\frac{z_1}{z_2}$. We shall therefore obtain a portion of the twisted curve in question if we introduce a factor of proportion ρ, and write the following equations:

$$\rho x_\nu = t_\nu(z_1, z_2), \tag{10}$$

and now consider $z_1 : z_2$ as current parameter. This clearly gives a rational, and therefore irreducible, twisted curve of the sixth order.* I say, now, *that this is half-regular, and therefore our twisted curve of the twelfth order supplies, besides* (10), *a second rational twisted curve of the sixth order, which is derived from* (10) *by an arbitrary odd permutation of the x's.*

To prove this we show, first, that the curve (10) actually admits the even collineations. This cannot indeed be otherwise, since the curve of the twelfth order remains unchanged for all the 120 collineations and $12 = 2 . 6$; but we will prove it directly. We allow $z_1 : z_2$, starting from any initial value, to vary continuously in such a way that it assumes successively all the 60 values which proceed from the said initial value by means of the 60 icosahedral substitutions. Then the point x—since we are concerned throughout with continuous variations—always moves on the same irreducible curve, while, at the same

* The formulæ (10) cannot represent some curve of lower order repeated, for we can calculate $z_1 : z_2$ rationally from the corresponding x_ν's.

time, as we know beforehand, the t_ν's have undergone at the end all the even permutations. The curve, therefore, is in fact transformed into itself by all the even collineations.

We prove, moreover, that our curve cannot admit further collineations, viz., if this was the case, then would Z (which in consequence of equation (6) can be represented as a symmetric function of the u_ν's) assume the same value not only in 60, but in 120 points of our curve of the sixth order, while yet to every value of Z corresponding to the icosahedral equation (9) only 60 values $z_1 : z_2$ belong.

With this our primary assertion is fully proved. We should evidently have been able to confirm this by only making use of the formulæ (9) and (10), and leaving aside the consideration of the sums of the powers and formula (7). In this way we will now discuss those curves which belong geometrically to what we previously called the *canonical resolvent* of the icosahedral equation (I, 4, § 12). We have there given a definition of the roots Y_ν, which we can here reproduce with the addition of a factor of proportion ρ, in the following way:

$$(11)\qquad \begin{aligned}\rho Y_\nu = {} & m \,.\, W_\nu(z_1, z_2) \,.\, T(z_1, z_2) \\ & + 12n \,.\, t_\nu(z_1, z_2) \,.\, W_\nu(z_1, z_2) \,.\, f^2(z_1, z_2):\end{aligned}$$

here t_ν is the given form of the sixth degree, f and T are the usual icosahedral forms, and W_ν is equal to the following expression:

$$(12)\qquad \begin{aligned}W_\nu = {} & -\epsilon^{4\nu}z_1^8 + \epsilon^{3\nu}z_1^7 z_2 - 7\epsilon^{2\nu}z_1^6 z_2^2 - 7\epsilon^{\nu}z_1^5 z_2^3 \\ & + 7\epsilon^{4\nu}z_1^3 z_2^5 - 7\epsilon^{3\nu}z_1^2 z_2^6 - \epsilon^{2\nu}z_1 z_2^7 - \epsilon^{\nu}z_2^8.\end{aligned}$$

If we now allow $z_1 : z_2$ to vary, the point Y in virtue of (11) traces out, as $m : n$ changes its value, an infinite number of rational curves of the 38th order, among which a curve of the eighth order for $n = 0$, and a curve of the fourteenth order for $m = 0$, are included.* *All these curves are half-regular.* They will therefore be accompanied by one of two curves of the same order, which arise from (11) by an arbitrary odd permutation of the Y_ν's. Only when taken jointly—in general, therefore, a curve of the 76th order—are the two curves the geometrical

* I leave for the time undiscussed whether or no other curves of the system suffer a reduction of order, and also the question on what—geometrically speaking—this reduction actually depends.

image of the individual canonical resolvent. We consider, moreover, *that all these curves are situated on the canonical surface;* for ΣY^2 is in general for the canonical resolvent equal to zero. The closer investigation of *how* these curves move on the canoni ' surface, and what relations exist between them and the linear generators of the canonical surface, will occupy us more in detail in the next chapter.

§ 5. Geometrical Conception of the Tschirnhausian Transformation.

In order now to make the Tschirnhausian transformation of the equations of the fifth degree accessible to our geometrical interpretation, we will, in correspondence with the condition (1), in consequence of which the sum of the roots of the equations in question must always vanish, introduce the following notation:

$$(13)\qquad x_\nu^{(1)} = x_\nu - \frac{s_1}{5},\ x_\nu^{(2)} = x_\nu^2 - \frac{s_2}{5},\ x_\nu^{(3)} = x_\nu^3 - \frac{s_3}{5},\ x_\nu^{(4)} = x_\nu^4 - \frac{s_4}{5},$$

(where, of course, $x_\nu^{(1)}$ is only written for x_ν for the sake of uniformity). Then the most general transformation which we will consider is this:

$$(14)\qquad y_\nu = p\,.\,x_\nu^{(1)} + q\,.\,x_\nu^{(2)} + r\,.\,x_\nu^{(3)} + s\,.\,x_\nu^{(4)},$$

understanding by p, q, r, s any magnitudes at first indeterminate.

We have hitherto only considered such expressions as are transformed into themselves for the particular fundamental permutations or linear transformations, and which are therefore *invariants* with respect to the transformation group. In a corresponding sense we could describe the expressions (13) as *covariants* of the x_ν's, inasmuch as they are permuted simultaneously with the x_ν's, and in like manner. I will not explain further here how we construct geometrically the covariant points $x^{(2)}$, $x^{(3)}$, $x^{(4)}$ from the given points $x = x^{(1)}$ in the most effective manner. On the other hand, I should like to call attention to the fact that, in virtue of (14), an arbitrary point y will be constructed from the four fundamental points $x^{(1)}$, $x^{(2)}$, $x^{(3)}$, $x^{(4)}$ by the help of proper multipliers, p, q, r, s, in just the same way as is usually done in the projective geometry (since the barycentric calculus of Möbius). *The p, q, r, s are therefore nothing else than new projective co-ordinates of the point y, which are related to covariants of x;* or, to express it in the more suggestive terms of modern phraseology: *the assertion* (14) *denotes that instead of the original co-ordinate system of x a typical co-ordinate system is introduced.** In the application of the Tschirnhausian transformation we are concerned with the problem so to determine p, q, r, s that the transformed equation in y which results may have any special properties with respect to the variability of its coefficients. Geometrically speaking, we must constrain the point y to move only on predetermined surfaces or curves. We will therefore write down the equations of these surfaces or curves, and see how we can find any system of values for p, q, r, s which satisfies these equations.

We have already given in § 2 of the previous chapter some elementary remarks on the problem here enunciated. Moreover, the distinctions which were just now developed in § 3 will here be of importance. For it is evidently sufficient, where the main surface or curve which we are considering is reducible, to write down the equation of only a single irreducible portion of the surface or curve. If m is the number of those of the 120 collineations by which the portion in question is transformed into itself, then the coefficients of those equations which we use for the expression of this portion in our new system of co-ordinates will so depend on $x_\nu^{(1)}$, $x_\nu^{(2)}$, &c., that they remain unaltered for the said m-permutations of the x's, and for these only. The coefficients will therefore only be symmetric functions of the x_ν's when we have to deal with regular figures, but two-valued functions (which, after adjunction of the square root of the discriminant, are rational) when half-regular figures are considered, &c.

It follows from this *that in the solution of equations of the fifth degree the Tschirnhausian transformation will be only of use in those cases where regular or half-regular figures are given.* For if we wanted to include irregular figures, we should first have to adjoin, merely for the purpose of constructing their typical

* *Cf.* Clebsch, "Theory of Binary Algebraic Forms," p. 300, &c.

equations, such functions of the x_ν's that there would no longer remain a real problem or one which demanded any but the most elementary processes. Here comes into play the somewhat incidental circumstance that the group of 60 even permutations of five things is simple, and therefore loses its more distinctive characteristics by any further adjunction.

§ 6. Special Applications of the Tschirnhausian Transformation.

In order to determine a point on a given surface or curve of the nth order, the readiest method in any case is that, for which an auxiliary equation of the nth degree will be required, of cutting the surface with a known *straight line*, or the curve with a *plane*. For the Tschirnhausian transformation as it is given by (14), this gives the following general lemma. We take two or even three sets of known magnitudes:

$$P_1, Q_1, R_1, S_1\,;\; P_2, Q_2, R_2, S_2\,;\; P_3, Q_3, R_3, S_3,$$

and then put either

(15) $$p=\rho_1 P_1+\rho_2 P_2,\; q=\rho_1 Q_1+\rho_2 Q_2,\; r=\rho_1 R_1+\rho_2 R_2,\; s=\rho_1 S_1+\rho_2 S_2,$$

or

(16) $$p=\rho_1 P_1+\rho_2 P_2+\rho_3 P_3, \text{ \&c.}$$

If we then introduce these values into the equation of the surface, or the equations of the curve, as the case may be, we obtain for $\rho_1 : \rho_2$ an equation, or for $\rho_1 : \rho_2 : \rho_3$ a system of equations of the nth order; each root of this equation or of this system of equations (as the case may be) gives us a Tschirnhausian transformation of the required properties. The irrationality which is thus required for the production of the transformation is evidently in general an *accessory* one. For there is no *a priori* reason why the discrimination of the n-points of intersection of an arbitrary straight line with the surface, or of a plane with a curve, should have anything to do with the distinction of the collineations which transform this surface or curve into themselves.

It need hardly be said that the general process thus described, practically speaking, does not take us far. If we tried to treat the different special cases (enumerated in §§ 3, 4) of equations of the fifth degree by its means, we should be brought at once after the first two cases to auxiliary equations of higher degree than the fifth. *We will therefore in the sequel only use our general process, or suppose it used, in order to transform the general equation of the fifth degree into a canonical equation.* In fact, we shall afterwards (in the fifth chapter) bring forward proof that in this special case the general process cannot be improved, inasmuch as it is in no way possible to get rid of the accessory square root which is introduced by our process. On the other hand, we shall succeed in all the other cases in finding more simple methods for the production of the transformation. These methods were partly touched upon in the developments of the preceding chapter; we add here a few supplementary remarks.

First, as regards Bring's transformation, we have stated already that it is possible, instead of the original system of equations of the sixth degree with which we have to deal, to substitute a sequence of quadratic equations and a cubic equation. We can now, in reliance on our geometrical method of representation, express this much more precisely. The theory is marshalled in detail as follows. We first of all transform the general equation of the fifth degree, in the way above described, into a canonical equation (where we employ a first square root and an accessory square root). *But then arises, geometrically, the important fact that through every point on the canonical surface pass two linear generators thereof, of which each meets Bring's curve in only three other points.* We shall therefore, in order to pass from an arbitrary point on the canonical surface to a point on Bring's curve, first employ another square root in order to define the generator passing through the point, and then, in fact, obtain an equation of the third degree, which determines the points of intersection of the chosen generator with Bring's curve. It has been already stated that we shall establish later on (in fact, in the third and next chapter) explicit formulæ for all the steps required for Bring's theory. We only observe here, therefore, what was for the most part passed over in laying down the theory, *that the second square root* (which defines the two generators of the principal surface) *is not an accessory one, but coincides with the square root of the discriminant of the equation of the fifth degree.* The irrationality which will be introduced by the cubic auxiliary equation is, on the contrary, again an accessory one; the cubic equation is also in Galois's sense general, *i.e.*, such as possesses a group of six permutations.

We discuss, moreover, the equations of the fifth degree established by Brioschi, which depend on the Jacobian equations of the sixth degree. By the existence of Kronecker's resolvent a method is indicated, as we remarked before, of transforming the general equations of the fifth degree into these special ones. In the first place, we have here to deal with the *diagonal equation* of the fifth degree. Our previous account shows that only the square root of the discriminant, and therefore in no way an accessory square root, is required in order to turn the general equation of the fifth degree into a diagonal equation. If we assume an accessory square root, we can ensure that $A=0$ in Kronecker's resolvent. The corresponding diagonal equation is then essentially identical with the equation of the u's which we just considered in § 4. The curve of the u's was of the twelfth order, or split up into two half-regular curves of the sixth order. Our general proposition would, therefore, for this also lead to an auxiliary equation of the sixth degree after adjunction of the square root of the discriminant. Nevertheless, as has just been stated, a single additional square root is sufficient.

§ 7. Geometrical Aspect of the Formation of Resolvents.

The algebraical principles of the construction of resolvents have already been thoroughly explained in I, 4 for arbitrary algebraical equations. Their specification for equations of the fifth degree needs in itself no corollary. If we return to this here, it is only to give a new application to our former remarks.

Let us agree, in the first place, that we will only introduce such rational functions of the x's:

$$\phi(x_0, x_1, x_2, x_3, x_4)$$

as roots of the equation, as are homogeneous in the x's; if we then multiply all the x's by the same factor λ (where the representative point, which we shall call x, remains unaltered), the ratios of those values:

$$\phi_0, \phi_1, \ldots \phi_{n-1},$$

which ϕ assumes in consequence of our permutations, are shown to be invariants in every case, and we can therefore, by denoting the ϕ's as homogeneous co-ordinates, interpret the formation of the resolvents in a geometrical way. This is a limitation which we make merely in favour of our geometrical interpretation; it has no deeper significance, and can hereafter be dispensed with.

Corresponding to the basis of analytical geometry, two possibilities now occur at the outset for the interpretation. Either we consider the introduction of the ϕ's as a mere *change of the system of co-ordinates*, or, in Plücker's sense, as *a change of the elements of space.* In the first case, the ϕ's appear directly as homogeneous, and in general curvilinear co-ordinates of a point, between which $(n-4)$ identities necessarily exist. In the second case, the ϕ's are primarily independent magnitudes, which we denote as the co-ordinates of any geometrical figure. The choice of this figure is only restricted by the condition that its co-ordinates, on the introduction of the 120 or 60 collineations of space which we are considering, experience just the same permutations as the ϕ's undergo as functions of the x's. Putting then the ϕ's equal to the said functions of the x's, *let us establish a covariant relation between such a figure and the point* x. The solution of the equation of the fifth degree by the formation of resolvents consists, therefore, in finding first, instead of the point x, another figure covariant to it, and then returning finally from this to the point x.

In what follows we shall for the most part keep to the second and more significant representation of the formation of resolvents, and indeed so much so, that we shall choose it at once as the starting-point of our further considerations. The simplest figures of space are, in respect of their exhibition by means of projective geometry, the point, the plane, and the straight line. We can consider in their order the resolvents which arise when we start from these very figures, using for our system of co-ordinates the simplest possible.

The consideration of *covariant points, related to the original pentahedron,* tells us, of course, nothing new, but leads back to the Tschirnhausian transformation already disposed of. We have only here to introduce p, q, r, s as invariants of the x's, *i.e.*, as symmetric functions of them, or as functions which remain unaltered for the sixty even permutations. The use of covariant planes is not more profitable, viz., if we consider, as we naturally may, as co-ordinates of the plane the coefficients u of its equation:

$$(17) \qquad u_0x_0 + u_1x_1 + u_2x_2 + u_3x_3 + u_4x_4 = 0,$$

where we suppose this equation so regulated by the help of $\Sigma x = 0$ that Σu is also always $=0$, then to every plane there belongs a covariant point with just the same co-ordinates. This is its *pole* with respect to the canonical surface $\Sigma x^2 = 0$. In fact, if $x'_0 \ldots x'_4$ are the co-ordinates of the pole (where $\Sigma x' = 0$), the equation of the polar plane is easily found to be:

$$(18) \qquad x_0'x_0 + x'_1x_1 + x'_2x_2 + x'_3x_3 + x'_4x_4 = 0,$$

and is therefore identical with (17) if we make the several u's equal to the x's respectively. Consequently the same five magnitudes can always be looked up either as point- or as plane-co-ordinates, and a special consideration of the plane as the element of space is useless.

Thus there remain as the simplest resolvents which we can consider those which start from *a straight line covariant to the point x.* Before I go further into this, I shall make* a few prefatory remarks concerning line co-ordinates in space and on the general principles of line geometry; first, because this matter, apart from the sphere of geometry proper, may still be little known, and also because we shall have to consider, instead of the usual tetrahedral co-ordinates, a pentahedron.

* See Plücker's "Neue Geometrie des Raumes, gegründet auf die Betrachtung der geraden Linie als Raumelement" (Leipzig, 1868–69); as well as the new edition of Salmon-Fiedler's "Analytical Geometry of Space."

§ 8. On Line Co-ordinates in Space.

The special principle of line co-ordinates in space, which we can retain as well by the use of pentahedral as of tetrahedral co-ordinates, was given by Grassmann as far back as 1844 in the first edition of his "Ausdehnungslehre" (Leipzig, Wigand).* Let X, Y be two points on the straight line, *then we consider as homogeneous line co-ordinates the entire set of binary determinants, which can be constructed with the co-ordinates of these points.* Let us take first as our foundation, keeping to the usual mode of representation, a co-ordinate tetrahedron. We then denote the co-ordinates of X, Y as follows:

$$X_1, X_2, X_3, X_4; \; Y_1, Y_2, Y_3, Y_4,$$

putting

(19) $$p_{ik} = X_i Y_k - Y_i X_k,$$

we have in the first place:

(20) $$p_{ik} = -p_{ki},$$

by means of which the twelve different p_{ik}'s which occur are reduced to six linearly independent ones, for which we choose, say, the following:

(21) $$p_{12}, \quad p_{13}, \quad p_{14}, \quad p_{34}, \quad p_{42}, \quad p_{23}.$$

Between these there then exists in addition the following easily-proved identity:

(22) $$P = p_{12}p_{34} + p_{13}p_{42} + p_{14}p_{23} = 0.$$

Two lines *intersect* when a bilinear relation obtains among their co-ordinates, which we can denote briefly as follows:

(23) $$\sum p'_{ik} \cdot \frac{\delta P}{\delta p_{ik}} = 0.$$

The summation has here to extend over the six combinations (21). This is clearly not the *general* linear equation for the p_{ik}'s, for the p'_{ik}'s are also subject to an identity of the form (22). Understanding by a_{ik} arbitrary magnitudes, and keeping

* Republished 1878.

to the table (21), we will write the general equation in question in the following form:

$$(24) \qquad \sum a_{ik} \cdot \frac{\delta P}{\delta p_{ik}} = 0.$$

The assemblage of the straight lines which satisfy an equation of this kind is what *Plücker* has called a *linear complex*, while *Möbius* in 1833 has discussed it more completely. We will not here concern ourselves with the geometrical properties of the linear complex any further. We will only add that we shall denote the coefficients a_{ik} as *co-ordinates of the linear complex*, where we may introduce, if we please, in accordance with formula (20), beside the a_{ik}'s, other symbols a_{ki}:

$$(25) \qquad a_{ik} = -a_{ki}.$$

If:

$$(26) \qquad a_{12}a_{34} + a_{13}a_{42} + a_{14}a_{23} = 0,$$

we can replace the a_{ik}'s by the p'_{ik}'s of the formula (23), the complex is then a *special* one, and consists evidently of all straight lines which intersect the fixed line p'. If we combine by addition two special complexes p', p'', and so construct:

$$a_{ik} = \lambda' p'_{ik} + \lambda'' p''_{ik},$$

we have, apart from particular cases, a general complex. *Every general complex can be obtained by adding together six given special complexes with the help of proper multipliers;* only the special complexes must be linearly independent, *i.e.*, they must not satisfy by their co-ordinates the same linear homogeneous equation. In this sense, in particular, the six straight lines are available which form the edges of a tetrahedron.

So much for the usual conventions of the line-geometry. If we now replace the co-ordinate tetrahedron by a pentahedron, the only modification is this: that the number of the co-ordinates appears to be increased, but, to meet this, new equations of condition occur. First as regards the point co-ordinates, we have for X, Y now, just as before:

$$X_0, X_1, X_2, X_3, X_4; \; Y_0, Y_1, Y_2, Y_3, Y_4,$$

with $\Sigma X = 0$, $\Sigma Y = 0$. *But then we have twenty determinants:*

(27) $$p_{ik} = X_i Y_k - Y_i X_k$$

to distinguish. We have again, of course:

(28) $$p_{ik} = -p_{ki};$$

but besides this, evidently:

(29) $$\sum_i p_{ik} = 0, \text{ or also } \sum_k p_{ik} = 0,$$

where the summation extends over those four values of i and k respectively, which are different from the corresponding k and i. Besides this, there exists the quadratic relation (22) in addition to the others which proceed from it by means of (28) and (29). Again, we can also speak of *co-ordinates of the linear complex.* There are twenty magnitudes a_{ik} which, while satisfying the linear relations (28), (29), are otherwise unrestricted variables. What was said with regard to the composition of general linear complexes out of special complexes remains valid. All these matters are so simple that we can now break off any further consideration of them.

§ 9. A Resolvent of the Twentieth Degree of Equations of the Fifth Degree.

Let us go back again to the considerations of § 7. We wished to consider those equations on which the pentahedral co-ordinates of a straight line in space depend. We can evidently, instead of these equations, at once take into consideration the more general ones by which the co-ordinates of an arbitrary linear complex are determined. *We thus obtain in general equations of the twentieth degree whose roots* a_{ik}, *in conformity to the formulæ* (28), (29), *are connected by the following linear relations:*

(30) $$a_{ik} = -a_{ki}, \quad \sum_i a_{ik} = 0, \quad \sum_k a_{ik} = 0.$$

A certain similarity between these equations and the Jacobian equations of the sixth degree (in so far as we regard the latter, as Herr Kronecker does, as equations of the twelfth degree for the $\sqrt{z}$'s) is from the very first unmistakable; we shall learn later on (in the fifth chapter) the intimate connection that actually exists in this respect.

Our business now is to make the magnitudes a_{ik} equal to proper functions of the x's, and to turn our equations of the twentieth degree into resolvents of the equation of the fifth degree. The plan is, as we expressed it in § 7, to connect the linear complex (whose co-ordinates are a_{ik}) with the point x as a covariant. We effect this in a simple manner if we adopt the methods of § 5. We have there constructed the $x^{(1)}$, $x^{(2)}$, $x^{(3)}$, $x^{(4)}$, as the simplest covariant points of the point x; we shall obtain the simplest covariant straight lines if we consider the lines which join these points. The co-ordinates p_{ik} of this line:

(31) $$p_{ik}^{l,m} = x_i^{(l)} x_k^{(m)} - x_i^{(m)} x_k^{(l)}$$

are linearly independent, for we have to do with the six edges of a tetrahedron. *Therefore we shall obtain the most general values of a_{ik} by combining these p_{ik}'s with the help of proper multipliers:*

(32) $$a_{ik} = \sum c^{l,m} \,.\, p_{ik}^{l,m}.$$

Here the $c^{l,m}$'s are to be introduced as symmetric or as two-valued functions of the x's, according as we choose to consider all the permutations of the x's, or only the positive permutations thereof; but otherwise they are to be chosen so that the law of homogeneity that we accorded is satisfied.

§ 10. Theory of the Surface of the Second Degree.

I conclude the present chapter with some remarks on the *institution of parameters* for the linear generators on surfaces of the second degree. The parameters in question are linear multipartite functions of the projective point co-ordinates.* We obtain them most simply by bringing the equation of the surface (as it is possible to do in an infinite number of ways) into the following form:

(33) $$X_1 X_4 + X_2 X_3 = 0.$$

If we put then, firstly, in correspondence with this equation:

* The introduction of this parameter is an equivalent, geometrically speaking, of the projective generation of the two families of ruled lines on the surface, which, for example, Steiner makes the basis of his considerations.

(34) $$\frac{X_1}{X_2} = \frac{X_3}{X_4} = \lambda,$$

secondly:

(35) $$\frac{X_1}{X_3} = -\frac{X_2}{X_4} = \mu;$$

λ remains constant when we move along a generator of one kind, which might be called the first, while μ remains constant when we proceed along a generator of the second kind. Therefore λ, μ are two numbers which are characteristic of the individual generators of the first or second kind, *i.e.*, they are parameters which can be used to distinguish the generators. Here we observe that each of the formulæ (34), (35), embraces two equations. We may therefore, without changing the meaning of λ, μ, generalise their definition somewhat. For λ, for example, by combining the two equations (34) with the help of arbitrary magnitudes ρ and σ, we can write:

(36) $$\lambda = \frac{-\rho X_1 + \sigma X_3}{+\rho X_2 + \sigma X_4}.$$

We succeed in making numerator and denominator of λ vanish together for an arbitrarily chosen generator of the second kind:

$$\mu = \frac{\sigma}{\rho}.$$

The generator chosen in this manner shall be called *the basis of the introduction of* λ.

We will now first consider the behaviour of λ, μ with respect to such space collineations as transform our surface into itself.* The collineations in question arrange themselves, as is known, into two kinds, according to their behaviour with regard to the generators of the surface: *either they transform each of the two systems of generators into itself, or they interchange the two systems.* In the first case, to each generator corresponds, in virtue of the presupposed collineation, one, and only one, generator λ'; and conversely, in the same way, to every μ corre-

* Consult, say, Bd. ix of the Math. Ann., p. 188, &c. The theorems introduced in the text are also often used in other departments of modern research. A more thorough proof would take us too far.

sponds one μ'. *Therefore we have, on function-theory principles, corresponding to such a collineation, formulæ of the following kind necessarily appertaining:*

$$(37) \qquad \lambda' = \frac{a\lambda + b}{c\lambda + d}, \quad \mu' = \frac{a'\mu + b'}{c'\mu + d'}.$$

In other cases λ' by analogy will be a linear function of μ; μ' such a function of λ. I do not stay to show that these propositions can also be reversed, and that therefore a corresponding space collineation is obtained if the formulæ (37) [or the corresponding ones in which λ and μ are interchanged] are written down quite arbitrarily.

We remark, moreover, that the λ, μ's *furnish a determination of co-ordinates for the points on our surface.** In fact, at every point one generator of the first kind and one of the second intersect, whose λ, μ we can transfer to the point. It is here to the purpose to replace λ by $\lambda_1 : \lambda_2$, μ by $\mu_1 : \mu_2$, to make them homogeneous. An algebraical equation:

$$(38) \qquad f(\lambda_1, \lambda_2; \mu_1, \mu_2) = 0,$$

homogeneous and of degree l in λ_1, λ_2, of degree m in μ_1, μ_2, then expresses a curve of the $(l+m)^{\text{th}}$ order lying on the surface, which intersects a generator of the first kind m times and one of the second kind l times. We can now combine (34), (35) in the following manner:

$$(39) \qquad X_1 : X_2 : X_3 : X_4 = \lambda_1\mu_1 : -\lambda_2\mu_1 : \lambda_1\mu_2 : \lambda_2\mu_2.$$

Introducing these values of the X's into the equation of a surface of the nth order:

$$(40) \qquad F(X_1, X_2, X_3, X_4) = 0,$$

we recognise that our surface of the second degree is intersected by (40) in a curve which, written in a form (38), is of the nth degree both in the λ's and μ's. Conversely, too, by means of formula (39), every curve (38) which is of equal degree in the λ, μ's can be represented as the complete inter-

* See Plücker in Crelle's Journal, vol. xxxvi (1847). The discussion of the curves (38) was undertaken in a systematic manner almost simultaneously by Mr. Cayley and by Chasles (1861); see Phil. Mag., vol. xxii, also Comptes Rendus, vol. liii.

section of the surface of the second degree with an accessory surface (40).*

We determine, finally, *the line co-ordinates* of the generator λ, μ, retaining the tetrahedron as laid down in (33). Putting first $\mu_1=0$, then $\mu_2=0$ in (39), we obtain for two points lying on the generator λ:

$$X_1 : X_2 : X_3 : X_4 = 0 : \ 0 : \lambda_1 : \lambda_2$$
$$Y_1 : Y_2 : Y_3 : Y_4 = \lambda_1 : -\lambda_2 : 0 \ : 0$$

respectively.

Hence we calculate by (19) for the p_{ik}'s which belong to them the following relative values:

(41) $p_{12}=0,\ p_{13}=\lambda_1^2,\ p_{14}=\lambda_1\lambda_2,\ p_{34}=0,\ p_{42}=\lambda_2^2,\ p_{23}=-\lambda_1\lambda_2.$

Analogously we get for the μ-generator:

(42) $p_{12}=-\mu_1^2,\ p_{13}=0,\ p_{14}=\mu_1\mu_2,\ p_{34}=\mu_2^2,\ p_{42}=0,\ p_{23}=\mu_1\mu_2.$

We now assume that the equation of a linear complex is introduced, which runs as follows:

$$\sum A_{ik} \cdot \frac{\delta P}{\delta p_{ik}} = 0.$$

By inserting herein the expressions (41), (42), we obtain the two following quadratic equations:

(43) $$A_{42}\lambda_1^2 + (A_{23} - A_{14})\,\lambda_1\lambda_2 + A_{13}\lambda_2^2 = 0,$$

(44) $$-A_{34}\mu_1^2 + (A_{23} + A_{14})\,\mu_1\mu_2 + A_{12}\mu_2^2 = 0.$$

Hence:

In general to a linear complex belong two, and only two, generators of each system.

But it may happen that one or other of these equations

* I might perhaps add one remark, which is not immediately connected with the text, but rather reverts to the developments of the first part, viz., this: that Riemann's interpretation of $x+iy$ on the sphere can be applied as a special case of the determination of λ, μ co-ordinates spoken of in the text, namely, since all the linear generators on the sphere are imaginary, two conjugate imaginary generators intersect in every real point thereof. If we now introduce λ, μ properly, and call the λ, which belongs to a real point on the sphere, $x+iy$, then the corresponding μ will be $x-iy$. *For fixing the real point, therefore, it suffices to give only one value,* $x+iy$, and this is just the method of Riemann, which, however, I cannot here work out in detail. *Cf.* Math. Annalen, Bd. ix, p. 189 (1875).

vanishes identically. This gives three linear conditions for the A_{ik}'s, so that three of these still remain arbitrary. Hence:

*The generators of the first and second kind on our surface belong each to a threefold linear family of linear complexes.**

I must pass over the actual establishment of the equations of these families.

* *Cf.* throughout Plücker's "Neue Geometrie des Raumes," &c.

CHAPTER III

THE CANONICAL EQUATIONS OF THE FIFTH DEGREE

§ 1. Notation—The Fundamental Lemma.

THE new chapter which we now begin is to form in every respect the centre of our developments. We treat of the canonical equations of the fifth degree and their simple relations to the icosahedron. Here we borrow from what precedes, especially from the Bring transformation, the one fundamental idea of *considering the rectilineal generators of the canonical surface.* I denote here, as I did there, the canonical equation of the fifth degree as follows:

$$(1) \qquad y^5 + 5\alpha y^2 + 5\beta y + \gamma = 0,$$

where the factors 5 for α and β respectively are applied for the sake of convenience. I will also communicate at the outset the value of the discriminant. Using the somewhat long formula which we frequently find* given for the discriminant of the general equation of the fifth degree, we have for (1):

$$(2) \qquad \Pi\ (y_i - y_k)^2 = 3125\nabla^2,$$

where ∇^2 is put for brevity in place of the following expression:

$$(3) \qquad \nabla^2 = 108\alpha^5\gamma - 135\alpha^4\beta^2 + 90\alpha^2\beta\gamma^2 - 320\alpha\beta^3\gamma + 256\beta^5 + \gamma^4.$$

We now at once proceed from the developments just given (in the concluding paragraph of the preceding chapter) by supposing the two different generators of the canonical surface to be denoted by parameters λ, μ. Let:

* *Cf. e.g., Faà di Bruno*, edited by *Walter*, "Enleitung in die Theorie der binären Formen" (Leipzig, 1881), p. 317.

$$(4) \qquad y_0, y_1, y_2, y_3, y_4$$

be the roots of (1) in a definite order. We then suppose those 60 generators λ and 60 generators μ constructed which contain one of the 60 points of the canonical surface, whose co-ordinates proceed from (4) by an even permutation of the y's. The λ, μ's are, as we know, linear fractional functions of the y's; the 60 values of λ or μ in question therefore depend on an equation of the 60th degree, which is a rational *resolvent* of our principal equation, and the coefficients of which are accordingly rational functions of the α, β, γ, ∇. *Now I assert*—and here we have the particular lemma required for our further developments—*that our resolvents of the* 60*th degree, for an appropriate introduction of the* λ, μ*'s, are necessarily icosahedral equations, and therefore will admit of being written without more ado:*

$$(5) \qquad \frac{H^3(\lambda)}{1728 f^5(\lambda)} = Z_1, \quad \frac{H^3(\mu)}{1728 f^5(\mu)} = Z_2,$$

where Z_1, Z_2 *alone depend on* α, β, γ, ∇.

The proof presents itself immediately on the grounds of our previous data. We have just divided the collineations of space which transform a surface of the second degree into itself into two parts, according as they transform the individual system of generators into itself, or interchange it with the other system. Now the canonical surface of the second degree passes into itself for the 120 collineations of space which correspond to the permutations of the y's. We will at first leave undetermined how the systems of generators of the surface behave towards *the totality* of these collineations. If all the collineations were not to transform the individual system of generators into itself, at all events half of them would necessarily do so. This half of our collineations must here necessarily form a group *per se*, and indeed a self-conjugate group in the main group; it can therefore only consist of the *even* collineations. *Hence, in any case*—and this is a first result—*the* 60 *even collineations have the property of transforming each of the two systems of generators of the canonical surface into itself.* We now recall that, in accordance with the formula just given in (37) [II, 2, § 10], the parameter λ, as also the parameter μ, experiences on its part a linear transformation for each collineation of this kind. *The* 60 *values*

λ, *which satisfy our resolvent of the 60th degree, therefore depend on each other as linear functions with constant coefficients (and similarly the corresponding values of* μ), or *the equations for* λ *and* μ *are transformed respectively into themselves by a group of 60 linear substitutions.* But hence the accuracy of our assertion follows immediately in virtue of the developments of I, 5, § 2, as soon as we add that the group of the linear transformations which λ or μ experiences is simply isomorphic with the group of even permutations of the y's. The unknowns λ, μ, which occur in the canonical forms (5), are here proper linear functions of the original parameters denoted by these letters; we will call them the *normal parameters,* not forgetting, however, that they can be chosen in sixty different ways in correspondence with the 60 linear transformations by which each of the equations (5) is transformed into itself.

Having thus proved our primary assertion, we can go a step farther in the same direction. I say first, again taking up the question just mooted, *that for each uneven collineation the two systems of generators of the canonical surface are necessarily interchanged,* namely, if the individual system were transformed into itself for the whole of the 120 collineations, a group of 120 linear substitutions of a variable would be given, on the grounds of the formula (37) just cited, which would be simply isomorphic with the group of 120 permutations of five things, which, however, by I. 5, § 2, is impossible. If, therefore, we have represented λ (the parameter of the generators of the first kind) in any way as a fractional linear function of the y's, we obtain a parameter μ of the generators of the second kind by subjecting the y's occurring in λ to any uneven permutation. *In particular, we obtain the sixty normal values of* μ *if we apply to one of the normal values of* λ *the whole of the uneven permutations of the* y's. For these uneven permutations the coefficients α, β, γ, of course remain unaltered, while ∇ changes its sign. *The magnitudes* Z_1, Z_2, *occurring in the equations* (5), *only differ, therefore, in the sign of* ∇. We can give this theorem another application by introducing the sixty points y', whose co-ordinates are derived from the scheme (4) by uneven permutations of the y's. *We have, namely, for the representation of the generators of the first and second kind which pass through these points, the following equations respectively:*

$$(6)\qquad \frac{H^3(\lambda)}{1728f^5(\lambda)}=Z_2,\quad \frac{H^3(\mu)}{1728f^5(\mu)}=Z_1.^*$$

§ 2. Determination of the Appropriate Parameter λ.

The formulæ which we will now establish for the normal λ are in themselves peculiarly simple and easy to verify. If I nevertheless devote some space to deducing it, it is because I again wish to derive each individual result from reflexions which involve no computation.

As the *generating operations* of the icosahedral group we have previously (I, 2, § 6) found the two following:

$$(7)\qquad \begin{cases} S: z'=\epsilon z, \\ T: z'=\dfrac{-(\epsilon-\epsilon^4)z+(\epsilon^2-\epsilon^3)}{(\epsilon^2-\epsilon^3)z+(\epsilon-\epsilon^4)}. \end{cases}$$

We saw, further (I, 4, § 10), that the octahedral forms t_ν are permuted as follows with respect to these substitutions:

$$(8)\qquad \begin{cases} S: t'_\nu=t_{\nu+1} \\ T: t_0'=t_0,\ t_1'=t_2,\ t'_2=t_1,\ t_3'=t_4,\ t_4'=t_3. \end{cases}$$

The same formulæ of permutation hold good for the roots

* I first gave the reasoning developed in the text (as well as the corresponding formulæ of the two following paragraphs) in two communications to the Erlanger Societät on November 13, 1876, and January 15, 1877 ["Weitere Mittheilungen über das Ikosaeder I, II"]. I will now append, besides, the case of the equations of the third and fourth degrees for comparison. Let us denote the three roots x of an equation of the third degree having $\Sigma x=0$, in accordance with what has gone before, on a straight line. Let us then denote an arbitrary point of this straight line in the usual way by a parameter λ; then λ, for the whole of the six permutations of the x's, experiences linear substitutions of the dihedral type, and satisfies, when properly prepared, a *dihedral equation of the sixth degree.*

For the equations of the fourth degree we transfer the geometrical representation to the plane, and add to the condition $\Sigma(x)=0$ the second one $\Sigma x^2=0$, confining ourselves therefore to "canonical equations." We again represent, in the usual manner, by a parameter λ the points of the selected conic. This parameter then undergoes linear substitutions for the whole of the twenty-four permutations of the x's, and therefore satisfies, when properly prepared, an *octahedral equation* (or, after adjunction of the square root of the discriminant of the equation of the fourth degree, a *tetrahedral equation*).

of the several resolvents of the fifth degree for the icosahedron which we there established (I, 4), and in particular—a point to which we shall soon return—for the roots of the canonical resolvent. We shall want to arrange our new formulæ so that they fit in with those there given as closely as possible. *We shall therefore so choose the normal* λ *from among the sixty values of the parameter which come under consideration, that it undergoes exactly the substitutions* (7), *if we subject the* y*'s to the two permutations indicated by* (8).

The value of λ is fixed hereby, but not so its form as a function of the y's. First, we have yet to decide which generator of the second kind we will make the basis of the introduction of λ in the sense previously explained (II, 2, § 10). Secondly, we can modify numerator and denominator of λ by addition of arbitrary multiples of Σy (which is identically $=0$). In both respects we will make definite conventions.

For each linear substitution of λ or μ two values of the variable, *i.e.*, two generators of the first or second kind respectively remain fixed. We consider now in particular the operation S, and make the basis of the introduction of λ one of the two generators of the second kind which remain fixed under its action. Let λ, on this supposition, $=\frac{p}{q}$, where p and q denote two linear functions of the y's. On effecting in p, q that permutation of the y's which is likewise indicated by S, p', q' arise. Here $p'=0$, $q'=0$, have by hypothesis the same straight line in common as $p=0$, $q=0$; therefore for any y:

$$p' = ap + bq + m \,.\, \Sigma y,$$
$$q' = cp + dq + n \,.\, \Sigma y,$$

but $\frac{p'}{q'}=\lambda'$ is, in accordance with formula (7), to be equal to $\epsilon\lambda$ for all points of the canonical surface; and the points of the canonical surface are not distinguished from the other points of space by any linear relation among the co-ordinates. Hence the foregoing equations are necessarily transformed into the more simple:

$$p' = \epsilon d \,.\, p + m \,.\, \Sigma y,$$
$$q' = d \,.\, q + n \,.\, \Sigma y,$$

where d, m, n are primarily unknown. We can modify these equations as follows:

$$p' + \frac{m}{\epsilon d - 1} \cdot \Sigma y = \epsilon d \left(p + \frac{m}{\epsilon d - 1} \cdot \Sigma y\right),$$

$$q' + \frac{n}{d-1} \cdot \Sigma y = d\left(q + \frac{n}{d-1} \cdot \Sigma y\right).$$

We shall now be able, without affecting the equation $\lambda = \frac{p}{q}$, to denote the expressions $p + \frac{m}{\epsilon d - 1} \cdot \Sigma y$, $q + \frac{n}{d-1} \cdot \Sigma y$, which occur, in a concise manner, by p, q. Then we have simply:

$$(9) \qquad \begin{cases} p' = \epsilon d \,.\, p, \\ q' = d \,.\, q. \end{cases}$$

The result of this reflexion is thus as follows: *we can put, and this in two ways* (since one of two generators of the second kind had to be chosen), *our* $\lambda = \frac{p}{q}$ *in such wise that, after application of the permutation* S *to the* y*'s, the equation* (9) *is identically true.*

Now, however, it is known (and, moreover, easy to prove) that, for the permutation S of any magnitudes y, no other linear functions of the y's alter only by a constant factor save multiples of the expressions of Lagrange:

$$(10) \qquad \begin{cases} p_1 = y_0 + \epsilon y_1 + \epsilon^2 y_2 + \epsilon^3 y_3 + \epsilon^4 y_4, \\ p_2 = y_0 + \epsilon^2 y_1 + \epsilon^4 y_2 + \epsilon y_3 + \epsilon^3 y_4, \\ p_3 = y_0 + \epsilon^3 y_1 + \epsilon y_2 + \epsilon^4 y_3 + \epsilon^2 y_4, \\ p_4 = y_0 + \epsilon^4 y_1 + \epsilon^3 y_2 + \epsilon^2 y_3 + \epsilon y_4, \end{cases}$$

with which Σy, as an expression which remains entirely unaltered, would also be associated were it not in our case identically zero. As regards the changes of the p_k's for the permutation S, $p'_k = \epsilon^{4k} \,.\, p_k$. Hence the only three expressions for λ which satisfy the relations (9) are the following:

$$(11) \qquad \lambda = c_1 \,.\, \frac{p_1}{p_2}, \quad \lambda = c_2 \,.\, \frac{p_2}{p_3}, \quad \lambda = c_3 \,.\, \frac{p_3}{p_4};$$

of these expressions, the first and the third are available, but the second must be rejected. It can be shown, namely, that the line of intersection of $p_1 = 0$ and $p_2 = 0$, as also that of $p_3 = 0$ and $p_4 = 0$, belong, in fact, to the canonical surface, but not the straight line $p_2 = 0$, $p_3 = 0$. This is best proved by introducing the p_k's in place of the y_ν's in the equation of the canonical surface. We have from (9), on joining thereto $\Sigma y = 0$:

(12) $$5y_\nu = \epsilon^{4\nu}p_1 + \epsilon^{3\nu}p_2 + \epsilon^{2\nu}p_3 + \epsilon^{\nu}p_4,$$

and therefore:

$$25\Sigma y^2 = 10(p_1p_4 + p_2p_3),$$

so that the equation of the principal surface, *relatively to the co-ordinate system of Lagrange*, will be the following:

(13) $$p_1p_4 + p_2p_3 = 0;$$

whence the accuracy of our statement follows directly. From (13) it follows, further, that:

$$\frac{p_1}{p_2} = -\frac{p_3}{p_4};$$

if, therefore, we put, corresponding to the first formula (11), $\lambda = c_1 \cdot \frac{p_1}{p_2}$, *we must also put it* $= -c_1 \cdot \frac{p_3}{p_4}$.

It will now only remain to determine the factor c which occurs here. We construct λ' by submitting the y's to the permutation T, according to formula (8), and then inserting it in the corresponding formula (7). The equation thus arising cannot be an identity, because the generator of the second kind, which we used in establishing the λ, does not remain fixed under the action of T. It must, however, be a valid equation if we take account of the relations $\Sigma y = 0$, $\Sigma y^2 = 0$. We obtain, on comparing the proper terms on both sides, the value -1 for c_1. *Hence our normal* λ *is determinate:*

(14) $$\lambda = -\frac{p_1}{p_2} = \frac{p_3}{p_4},$$

in exact agreement with the value which we had adopted for the parameter λ in the concluding paragraph of the preceding chapter (formula (34)).*

* If we wish to establish in a similar manner for the equations of the third and fourth degrees, which we just mentioned, the roots of the dihedral equation and octahedral equation respectively, we obtain accordingly:

$$\lambda = \frac{x_0 + \alpha x_1 + \alpha^2 x_2}{x_0 + \alpha^2 x_1 + \alpha x_2}, \text{ and } \lambda = \frac{\sqrt{2}(x_0 + ix_1 + i^2x_2 + i^3x_3)}{+(x_0 + i^2x_1 + i^4x_2 + i^6x_3)} = \frac{-(x_0 + i^2x_1 + i^4x_2 + i^6x_3)}{\sqrt{2}(x_0 + i^3x_1 + i^6x_2 + i^9x_3)}$$

where $\alpha^3 = i^4 = 1$, so that, therefore, the quotients of the expressions of Lagrange are here also introduced.

§ 3. Determination of the Parameter μ.

The normal parameter μ, which we had in view for the generators of the second kind, was to proceed from the parameter λ by means of an uneven permutation of the y's. We satisfy this requirement if we take, again in agreement with the concluding paragraph of the preceding chapter (formula (35)):

$$\mu = -\frac{p_2}{p_4} = \frac{p_1}{p_3}. \tag{15}$$

In fact, this value results from (14) if we replace y_0, y_1, y_2, y_3, y_4 by y_0, y_3, y_1, y_4, y_2 respectively, and therefore permute y_1, y_3, y_4, y_2 cyclically.

But now we can evidently deduce the formula (15) from (14) in yet another way, viz., by replacing in (14) ϵ by ϵ^2 throughout. This change is then, of course, carried over to the substitutions S, T (7), and the icosahedral substitutions arising from them. *The substitutions, therefore, which μ and λ undergo for the even permutations of the y's, though by no means identical individually, are so, at any rate, in their totality; or rather we derive the one set from the others by changing ϵ into ϵ^2 throughout,* a theorem which is fundamental in what follows. In agreement with this we obtain, on applying to the μ the operation mentioned, not λ again, say, but $-\frac{1}{\lambda}$. This is that value which arises from λ in virtue of the icosahedral substitution denoted previously by U. To it corresponds the simultaneous interchange of y_1 with y_4, and of y_2 with y_3.

We will further adopt the formula (39) of II, 2, § 10. In virtue of this we now have, on replacing λ by $\lambda_1 : \lambda_2$, μ by $\mu_1 : \mu_2$:

$$p_1 : p_2 : p_3 : p_4 = \lambda_1\mu_1 : -\lambda_2\mu_1 : \lambda_1\mu_2 : \lambda_2\mu_2, \tag{16}$$

or, on introducing a proportion-factor ρ:

$$\rho y_\nu = \epsilon^{4\nu} \,.\, \lambda_1\mu_1 - \epsilon^{3\nu} \,.\, \lambda_2\mu_1 + \epsilon^{2\nu} \,.\, \lambda_1\mu_2 + \epsilon^{\nu} \,.\, \lambda_2\mu_2. \tag{17}$$

§ 4. The Canonical Resolvent of the Icosahedral Equation.

Having found the normal parameters λ, μ for any canonical equation of the fifth degree, we will apply our formulæ in particular to the canonical resolvent of the fifth degree, which we previously constructed, IV, 1, § 12, by supposing any icosahedral equation:

$$(18) \qquad \frac{H^3(z_1, z_2)}{1728 f^5(z_1, z_2)} = Z$$

to be given. We obtain in this manner a peculiarly simple result, which is of the greatest importance for the further progress of our development.

The canonical resolvent was defined by the formulæ:

$$(19) \qquad Y_\nu = m \cdot v_\nu + n \cdot u_\nu v_\nu,$$

where

$$(20) \qquad u_\nu = \frac{12 f^2 \cdot t_\nu}{T}, \quad v_\nu = \frac{12 f \cdot W_\nu}{H},$$

understanding by f, H, T, the ground-forms of the icosahedron, by t_ν, W_ν, the oft-mentioned forms of the sixth and eighth degrees. Let us now consider that we can write W_ν and $t_\nu W_\nu$ in the following manner:

$$(21) \qquad \begin{aligned} W_\nu = &(\epsilon^{4\nu} z_1 - \epsilon^{3\nu} z_2)(-z_1^7 + 7 z_1^2 z_2^5) \\ &+ (\epsilon^{2\nu} z_1 + \epsilon^{\nu} z_2)(-7 z_1^5 z_2^2 - z_2^7), \end{aligned}$$

$$(22) \qquad \begin{aligned} t_\nu W_\nu = &(\epsilon^{4\nu} z_1 - \epsilon^{3\nu} z_2)(-26 z_1^{10} z_2^3 + 39 z_1^5 z_2^8 + z_2^{13}) \\ &+ (\epsilon^{4\nu} z_1 + \epsilon^{\nu} z_2)(-z_1^{13} + 39 z_1^8 z_2^5 + 26 z_1^3 z_2^{10}). \end{aligned}$$

Hence Y_ν in formula (19) assumes the following form:

$$(23) \qquad Y_\nu = (\epsilon^{4\nu} z_1 - \epsilon^{3\nu} z_2) R + (\epsilon^{2\nu} z_1 + \epsilon^{\nu} z_2) S,$$

where R, S are linear functions of m, n. The expressions of Lagrange (which we here also denote by capital letters) become therefore:

$$(24) \qquad \begin{cases} P_1 = 5 z_1 \cdot R, & P_3 = 5 z_1 \cdot S, \\ P_2 = -5 z_2 \cdot R, & P_4 = 5 z_2 \cdot S. \end{cases}$$

Hence we get simply:

$$\lambda = \frac{z_1}{z_2}, \quad \mu = \frac{R}{S}. \tag{25}$$

We have therefore at the outset: *The parameter λ is identical with the unknown $z_1 : z_2$ of the original icosahedral equation*, or expressed geometrically: *The point*

$$y_\nu = m \,.\, v_\nu(\lambda) + n \,.\, u_\nu(\lambda) \,.\, v_\nu(\lambda) \tag{26}$$

lies on a generator of the first kind, whatever m and n may denote. If we here consider λ as a variable magnitude, the point y_ν, as we saw in the fourth paragraph of the preceding chapter, traverses a half-regular rational curve, which is in general of the 38th order. For the proof of this, we had replaced the formula (26) by the following (a proportion-factor ρ being introduced):

$$\begin{aligned} \rho y_\nu = {} & m \,.\, W_\nu(\lambda_1, \lambda_2) \,.\, T(\lambda_1, \lambda_2) \\ & + 12n \,.\, t_\nu(\lambda_1, \lambda_2) \,.\, W_\nu(\lambda_1, \lambda_2) \,.\, f^2(\lambda_1, \lambda_2). \end{aligned} \tag{27}$$

We now recognise, first, we may cursorily remark, the reason (geometrically speaking), why the order of the curve thus obtained can sink to 14 for $m = 0$, and to 8 for $n = 0$. *It is because, in the first case, the aggregate of the* 12 *generators of the first kind $f(\lambda_1, \lambda_2) = 0$, counted twice, is separated from the general curve of the* 38*th order; in the second case the aggregate $T(\lambda_1, \lambda_2) = 0$, counted once.* But we have now, besides, the following theorems for our curves (27). *We find that our curves meet the generators of the first kind only once, and therefore the generators of the second kind* 37 *times.* In fact, we have for every generator λ by (27) only one point of the curve. We find, moreover, *that through every point of a generator λ only one curve* (27) *passes, so that the canonical surface is covered by the family of curves* (27) *exactly once.* The individual points of the generator λ, viz., are given by the corresponding μ, which determines the generator of the second kind which passes through the point. But if we suppose λ, μ in (25) to be known, the corresponding $m : n$ is computed *linearly*.

We append to this two further remarks which will be useful later on. First, as regards m and n, we can compute these linearly from the y_ν's previously given in accordance with formula (26), not merely relatively, but determining their absolute values. These formulæ are not altered if we permute the y_ν's evenly in any way. For by the agency of the icosahedral substitutions of λ, the $u_\nu(\lambda)$'s, $v_\nu(\lambda)$'s occurring on the right-hand side always undergo the same even permutations as the y_ν's placed on the left side. *The m, n therefore depend rationally on the y_ν's in such wise that they remain unaltered for even permutations of the y_ν's;* or, to express it otherwise, *the m, n are capable of rational representation as functions of the given magnitudes* $\alpha, \beta, \gamma, \nabla$. We consider further the relation between λ, μ which is furnished by the formula (25). If we subject λ to any of the icosahedral substitutions, the μ, inasmuch as it depends on the corresponding Y_ν's (exactly in the way we saw in the preceding paragraph), undergoes other icosahedral substitutions which proceed from the given ones by changing ϵ into ϵ^2. Following the terminology which was introduced in this connection by *Herr Gordan*, we will describe the changes of μ as contragredient to the changes of λ. *The formulæ* (25) *provide us with infinitely many rational functions of λ which, in this sense, are contragrediently related to λ.**

§ 5. Solution of the Canonical Equations of the Fifth Degree.

We have already, in §§ 1, 2, given the means of reducing the solution of the canonical equations of the fifth degree to an icosahedral equation:

$$\frac{H^3(\lambda)}{1728 f^5(\lambda)} = Z_1, \tag{28}$$

by determining λ as a functon of the y's. If we now wish to express conversely the y_ν's by means of the individual root λ, we can evidently employ the equation (26). I will now write it so that m, n are provided with an index 1, so that the con-

* The theorems proved in the text, as well as the principles for the solution of the canonical equations of the fifth degree to be developed immediately, were brought before the Erlangen Society by Herr Gordan and myself simultaneously on the 21st of May 1877. Herr Gordan there started from essentially different points of view from those to which we afterwards return. My own exposition, too, was in some measure different from that now given in the text, and in many respects less simple. *Cf.* here throughout my comprehensive memoir, "Weitere Untersuchungen über das Ikosaeder," in Bd. xii of the Mathematische Annalen (August 1877).

nection of our formula with the icosahedral equation (28) may be evident. We have then:

$$y_\nu = m_1 \,.\, v_\nu(\lambda) + n_1 \,.\, u_\nu(\lambda) \,.\, v_\nu(\lambda); \tag{29}$$

when we afterwards consider μ instead of λ, Z_1, m_1, n_1 will have to be simultaneously transformed into Z_2, m_2, n_2. *In order that the solution of the canonical equation by the help of the icosahedral equation may be complete, we have evidently only to further determine the* Z_1, m_1, n_1 *as rational functions of the magnitudes* α, β, γ, ∇ *previously given.*

We shall see later on how the calculation thus required can be carried out *a priori*. In the meantime, let us follow a much more elementary method. We have in I, 4, § 12, explicitly computed the canonical resolvent of the icosahedral equation by considering Z, m, n, as arbitrary magnitudes, and in § 14 have given the corresponding square root of the discriminant. It now follows from the considerations of the preceding paragraph that every canonical equation of the fifth degree, after a fixed value has been determined for ∇, admits of being put, in one way only, into the form of the canonical resolvent. *We shall therefore be able to determine* Z_1, m_1, n_1 *in a rational manner by simply comparing the coefficients of the general canonical resolvent and the square root of its discriminant with the coefficients* α, β, γ, *of the given canonical equation* (1) *and the adjoined value of the corresponding* ∇. We will here always define ∇, as we did in I, 4, § 14, in the following way:

$$25\sqrt{5} \,.\, \nabla = \prod_{\nu<\nu'} (y_\nu - y_{\nu'}), \tag{30}$$

which is reconcilable with the formulæ (2) and (3) of the present chapter. Comparing now, first, only the two sets of coefficients, we obtain:

$$\tag{31} \left\{ \begin{aligned} Z \,.\, \alpha &= 8m^3 + 12m^2n + \frac{6mn^2 + n^3}{1-Z}, \\ \frac{Z \,.\, \beta}{3} &= -4m^4 + \frac{6m^2n^2 + 4mn^3}{1-Z} + \frac{3n^4}{4(1-Z)^2}, \\ \frac{Z \,.\, \gamma}{3} &= 48m^5 - \frac{40m^3n^2}{1-Z} + \frac{15mn^4 + 4n^5}{(1-Z)^2}. \end{aligned} \right.$$

I have here at the outset written Z, m, n, instead of Z_1, m_1, n_1, because Z_2, m_2, n_2 satisfy these equations equally well.

The further computation now takes the following form.* From the first of the equations (31) we obtain:

$$(32) \qquad \frac{n^2}{1-Z} = \frac{12\beta m + \gamma}{12\alpha}$$

On the other hand we form:

$$(33) \qquad \begin{cases} -m\gamma + \dfrac{n^2\beta}{1-Z} = -\dfrac{9}{4Z}\left(4m^2 - \dfrac{n^2}{1-Z}\right)^3, \\ \alpha^2 - \dfrac{4}{81} \cdot \dfrac{1-Z}{n^2} \cdot (3m\alpha + 2\beta)^2 = \dfrac{1}{Z}\left(4m^2 - \dfrac{n^2}{1-Z}\right)^3, \end{cases}$$

and hence:

$$\alpha^2 - \frac{4}{18} \cdot \frac{1-Z}{n^2} \cdot (3m\alpha + 2\beta)^2 = \frac{4}{9}\left(m\gamma - \frac{n^2\beta}{1-Z}\right).$$

We need only introduce here the value (32) of $\frac{n^2}{1-Z}$, in order to obtain for m a quadratic equation. If we rearrange this by multiplying up by the denominator, we have:

$$(34) \qquad 16m^2(\alpha^4 - \beta^3 + \alpha\beta\gamma) - \frac{4}{3}m(11\alpha^3\beta + 2\beta^2\gamma - \alpha\gamma^2) + \frac{1}{9}(64\alpha^2\beta^2 - 27\alpha^3\gamma - \beta\gamma^2) = 0.$$

On solving this we find:

$$(35) \qquad m = \frac{(11\alpha^3\beta + 2\beta^2\gamma - \alpha\gamma^2) \pm \alpha\nabla}{24(\alpha^4 - \beta^3 + \alpha\beta\gamma)},$$

where ∇^2 exactly agrees with (3) (as we may verify), and the sign $\pm$ remains for the time, of course, undetermined. With this value of m the other unknowns are at once determined too. First, as regards the value of Z, it is sufficient to introduce the value of $\frac{n^2}{1-Z}$ from (32) into the first of the equations (33); we thus find:

$$(36) \qquad Z = \frac{(48\alpha m^2 - 12\beta m - \gamma)^3}{64\alpha^2[12(\alpha\gamma - \beta^2)m - \beta\gamma]}.$$

* I borrow the process of elimination used in the text from a lecture of Gordan's in the winter 1880–81. Herr Kiepert has also similarly employed the comparison with the canonical resolvent ("Auflösung der Gleichungen fünften Grades") in the Göttinger Nachrichten of July 17, 1878, or Borchardt's Journal, t. 87 (1879).

We obtain n in a corresponding manner if we write the first of the equations (31) as follows:

$$\left(12m^2+\frac{n^2}{1-Z}\right)n=aZ-8m^3-6m\cdot\frac{n^2}{1-Z},$$

and now consider m and Z as known. The final formula is:

$$n=-\frac{96am^3+72\beta m^2+6\gamma m-12a^2Z}{144am^2+12\beta m+\gamma}. \tag{37}$$

In order now to determine the sign of ∇ in (35), and therefore in (36) and (37), in a way corresponding to the priority of the λ and the notation m_1, n_1, Z_1, let us compare (30) with the difference-product of the principal resolvent previously noted. It here suffices to consider a special case. We take, say, $m=1$, $n=0$, in the general principal resolvent, and therefore have, in consequence of the formula (31):

$$\alpha=\frac{8}{Z},\quad \beta=-\frac{12}{Z},\quad \gamma=\frac{144}{Z}.$$

At the same time we obtain by I, 4, § 14:

$$\prod_{\nu<\nu'}(Y_\nu-Y_{\nu'})=-25\sqrt{5}\frac{12^4(1-Z)}{Z^3},$$

and thus by formula (30):

$$\nabla=-\frac{12^4(1-Z)}{Z^3}.$$

Now, by formula (35), the m becomes in this case:

$$\frac{-11\cdot2^8-3\cdot12^3\cdot Z\pm4\cdot12^3(Z-1)}{2^{12}-7\cdot12^3\cdot Z};$$

therefore if, as we assumed, m is to be equal to 1, we have to apply the lower sign in (35).

Thus we have in general:

$$m_1=\frac{(11\alpha^3\beta+2\beta^2\gamma-\alpha\gamma^2)-\alpha\nabla}{24\,(\alpha^4-\beta^3+\alpha\beta\gamma)}, \tag{38}$$

and hence by (36), (37), *the* Z_1 *and* n_1. The corresponding values of m_2, Z_2, n_2 proceed from this by reversing the sign of ∇ throughout.

§ 6. Gordan's Process.

The method just developed for the computation of m_1, n_1, Z_1 has the advantage of working throughout on elementary lines, and with the use of results previously deduced. It cannot, however, be denied that a certain amount of skill, though of a very simple kind, is required to introduce the right combinations of the equations (31), and that, therefore, this method does not fit in well with the mode of exposition which we have otherwise maintained, in which we have endeavoured always to see before commencing the results of our computations in their main outlines. I will therefore briefly go into the features of the computation originally given by *Herr Gordan*, and this the more readily because certain other aspects are connected with it which are of use for our main conception of the problem of solution.* Let us first make clear the difficulties which oppose a direct computation of the magnitudes Z_1, m_1, n_1. We had, for example, the defining equation:

$$Z_1 = \frac{H^3(\lambda)}{1728 f^5(\lambda)},$$

where we may substitute for λ the one value:

$$\lambda = -\frac{p_1}{p_2}.$$

Then we have in Z_1 a rational function of the five roots $y_0 \ldots y_4$ before us which remains unaltered for all even permutations of the y's. But now this latter only occurs because the y's are connected by the equations of condition $\Sigma y = 0$, $\Sigma y^2 = 0$; it does not occur if we consider the y's as arbitrarily variable magnitudes. To express it otherwise, Z_1 is for the even permutations of the y's *actually*, though not *formally*, invariant. Now all rules which we meet with in the usual expositions on the computation of symmetric functions, &c., relate to functions of formal symmetry; these rules are, therefore, not immediately available for our purpose.

* *Cf.*, besides the note just mentioned, a communication of Gordan's to the Naturforscherversammlung at München (Sept. 1877), as well as the larger memoir, "Ueber die Auflösung der Gleichungen vom fünften Grade," in Bd. xiii of Math. Annalen (January 1878).

Herr Gordan surmounts this difficulty by satisfying the equations of condition $\Sigma y = 0$, $\Sigma y^2 = 0$, in a general manner by functions of independent magnitudes. He then has henceforth to do altogether with functions of *independent variables*, and can establish for them an algorithm which is analogous to a certain extent to the process already mentioned relating to symmetric functions.

The independent variables from which Herr Gordan starts are essentially none other than the homogeneous parameters λ_1, λ_2, and μ_1, μ_2. We have above already expressed the ratios of the p_k's, and, on the other hand, the ratios of the y_ν's in terms of these magnitudes [formulæ (16), (17)]. Herr Gordan renders the formulæ in question concise by supposing the absolute values of the λ, μ's determined appropriately, and writing accordingly as follows:—

(39) $$p_1 = 5\lambda_1\mu_1,\ p_2 = -5\lambda_2\mu_1,\ p_3 = 5\lambda_1\mu_2,\ p_4 = 5\lambda_2\mu_2,$$

whereupon y_ν becomes equal to the following expression:

(40) $$y_\nu = \epsilon^{4\nu} \,.\, \lambda_1\mu_1 - \epsilon^{3\nu} \,.\, \lambda_2\mu_1 + \epsilon^{2\nu} \,.\, \lambda_1\mu_2 + \epsilon^{\nu} \,.\, \lambda_2\mu_2.$$

Before going further in the description of Gordan's process, we will express all the magnitudes, given and required, by the λ, μ's thus introduced. I first bring together the formulæ for the coefficients α, β, γ of the proposed equation of the fifth degree and the corresponding ∇. We have:

(41) $$\alpha = -\frac{\Sigma y^3}{15} = -\lambda_1^3\mu_1^2\mu_2 - \lambda_1^2\lambda_2\mu_2^3 - \lambda_1\lambda_2^2\mu_1^3 + \lambda_2^3\mu_1\mu_2^2,$$

(42) $$\beta = -\frac{\Sigma y^4}{20} = -\lambda_1^4\mu_1\mu_2^3 + \lambda_1^3\lambda_2\mu_1^4 + 3\lambda_1^2\lambda_2^2\mu_1^2\mu_2^2 - \lambda_1\lambda_2^3\mu_2^4 + \lambda_2^4\mu_1^3\mu_2,$$

(43) $$\gamma = -\frac{\Sigma y^5}{5} = -\lambda_1^5(\mu_1^5 + \mu_2^5) + 10\lambda_1^4\lambda_2\mu_1^3\mu_2^2 - 10\lambda_1^3\lambda_2^2\mu_1\mu_2^4 - 10\lambda_1^2\lambda_2^3\mu_1^4\mu_2 - 10\lambda_1\lambda_2^4\mu_1^2\mu_2^3 + \lambda_2^5(\mu_1^5 - \mu_2^5),$$

(44) $$\nabla = \frac{\prod_{\nu<\nu'}(y_\nu - y_{\nu'})}{25\sqrt{5}}$$

$$= \lambda_1^{10}(\mu_1^{10} + 11\mu_1^5\mu_2^5 - \mu_2^{10}) + \lambda_2^{10}(-\mu_1^{10} - 11\mu_1^5\mu_2^5 + \mu_2^{10})$$
$$+ \lambda_1^9\lambda_2(25\mu_1^8\mu_2^2 - 50\mu_1^3\mu_2^7) + \lambda_1\lambda_2^9(-50\mu_1^7\mu_2^3 - 25\mu_1^2\mu_2^8)$$
$$+ \lambda_1^8\lambda_2^2(-75\mu_1^6\mu_2^4 + 25\mu_1\mu_2^9) + \lambda_1^2\lambda_2^8(-25\mu_1^9\mu_2 - 75\mu_1^4\mu_2^6)$$
$$+ \lambda_1^7\lambda_2^3(-50\mu_1^9\mu_2 - 150\mu_1^4\mu_2^6) + \lambda_1^3\lambda_2^7(+150\mu_1^6\mu_2^4 - 50\mu_1\mu_2^9)$$
$$+ \lambda_1^6\lambda_2^4(150\mu_1^7\mu_2^3 + 75\mu_1^2\mu_2^8) + \lambda_1^4\lambda_2^6(+75\mu_1^8\mu_2^2 - 150\mu_1^3\mu_2^7)$$
$$+ \lambda_1^5\lambda_2^5(11\mu_1^{10} - 504\mu_1^5\mu_2^5 - 11\mu_2^{10}).$$

Of the magnitudes *required*, Z_1 is known to us immediately as a function of the λ's:*

$$(45) \qquad Z_1 = \frac{H^3(\lambda_1, \lambda_2)}{1728 f^5(\lambda_1, \lambda_2)},$$

but the m_1, n_1 also readily admit of representation in terms of λ, μ. If we introduce, viz., into the defining equations:

$$y_\nu = m_1 \cdot v_\nu(\lambda_1, \lambda_2) + n_1 \cdot u_\nu(\lambda_1, \lambda_2) \cdot v_\nu(\lambda_1, \lambda_2)$$

or:

$$y_\nu = 12 m_1 \cdot \frac{f(\lambda_1, \lambda_2) \cdot W_\nu(\lambda_1, \lambda_2)}{H(\lambda_1, \lambda_2)} + 144 n_1 \cdot \frac{f^3(\lambda_1, \lambda_2) \cdot t_\nu(\lambda_1, \lambda_2) \cdot W_\nu(\lambda_1, \lambda_2)}{H(\lambda_1, \lambda_2) \cdot T(\lambda_1, \lambda_2)}$$

the values (40) for the y_ν's, we get on solution:

$$(46) \qquad m_1 = \frac{M_1}{12 f(\lambda_1, \lambda_2)}, \quad n_1 = \frac{N_1 \cdot T(\lambda_1, \lambda_2)}{144 \cdot f^3(\lambda_1, \lambda_2)},$$

where M_1, N_1, denote the following two forms, *linear* in μ_1, μ_2:

$$(47) \qquad M_1 = \left\{ \begin{array}{l} \mu_1(\lambda_1^{13} - 39\lambda_1^8\lambda_2^5 - 26\lambda_1^3\lambda_2^{10}) \\ -\mu_2(26\lambda_1^{10}\lambda_2^3 - 39\lambda_1^5\lambda_2^8 - \lambda_2^{13}) \end{array} \right.$$

$$(48) \qquad N_1 = \mu_1(7\lambda_1^5\lambda_2^2 + \lambda_2^7) + \mu_2(-\lambda_1^7 + 7\lambda_1^2\lambda_2^5).^*$$

We have now to represent the magnitudes Z_1, m_1, n_1, rationally in terms of α, β, γ, ∇, *on the basis of the formulæ* (41)—(48) *now given.*†

* The magnitudes Z_2, m_2, n_2, which are associated with Z_1, m_1, n_1, are omitted for the sake of brevity.

† As a verification of the expressions furnished for M_1, N_1, we may observe that the determinant

$$\begin{vmatrix} \frac{\delta M_1}{\delta \mu_1} & \frac{\delta M_1}{\delta \mu_2} \\ \frac{\delta N_1}{\delta \mu_1} & \frac{\delta N_1}{\delta \mu_2} \end{vmatrix}$$

is simply equal to $H(\lambda_1, \lambda_2)$.

§ 7. Substitutions of the λ, μ's—Invariant Forms.

We must now become acquainted with the changes to which λ_1, λ_2, μ_1, μ_2 are subject when the y's are permuted. These changes are not, however, *per se*, completely determined. For of the four magnitudes λ, μ, one is superfluous, even if we take account of the absolute values of the y_ν's. We found above that, for the even permutations of the y_ν's which we denoted by S and T, $\frac{\lambda_1}{\lambda_2}$ undergoes the icosahedral substitutions so named, while $\frac{\mu_1}{\mu_2}$ is subjected to substitutions which are derived from these by transforming ϵ into ϵ^2. We further remarked that $\frac{\mu_1}{\mu_2}$ proceeds from $\frac{\lambda_1}{\lambda_2}$ by the cyclic permutation (y_1, y_3, y_4, y_2), and that by a repetition of this operation $-\frac{\lambda_2}{\lambda_1}$ is derived from $\frac{\mu_1}{\mu_2}$. *On the basis of these theorems we shall now define for λ_1, λ_2, μ_1, μ_2, homogeneous linear substitutions of determinant* 1, *in such wise that conversely from it, in virtue of* (40), *the proper permutations of the y_ν's follow.* To this end let us first put, employing homogeneous icosahedral substitutions of determinant 1:

$$(49) \qquad S: \lambda'_1 = \epsilon^3\lambda_1,\ \lambda'_2 = \epsilon^2\lambda_2;\ \mu'_1 = \epsilon\mu_1,\ \mu_2' = \epsilon^4\mu_2;$$

$$(50) \qquad T: \begin{cases} \sqrt{5}\,.\,\lambda'_1 = -(\epsilon - \epsilon^4)\,\lambda_1 + (\epsilon^2 - \epsilon^3)\,\lambda_2, \\ \sqrt{5}\,.\,\lambda'_2 = (\epsilon^2 - \epsilon^3)\,\lambda_1 + (\epsilon - \epsilon^4)\,\lambda_2; \\ \sqrt{5}\,.\,\mu'_1 = (\epsilon^2 - \epsilon^3)\,\mu_1 + (\epsilon - \epsilon^4)\,\mu_2, \\ \sqrt{5}\,.\,\mu'_2 = (\epsilon - \epsilon^4)\,\mu_1 - (\epsilon^2 - \epsilon^3)\,\mu_2, \end{cases}$$

where the formulæ for the μ's again proceed from those for the λ's on replacing ϵ by ϵ^2.*

Applying these substitutions to (40), there follows in fact necessarily:

$$S: y_\nu' = y_{\nu+1},$$
$$T: y_0' = y_0,\ y_1' = y_2,\ y_2', = y,\ y_3' = y_4,\ y_4' = y_3.$$

* Here $\sqrt{5} = \epsilon + \epsilon^4 - \epsilon^2 - \epsilon^3$, as must not be overlooked, changes its sign.

Here the permutations of the y's which arise by composition of S and T are, of course, only hemihedrically isomorphic with the corresponding substitutions of the λ, μ's: there are 120 substitutions of the λ, μ's, and only 60 permutations of the y's. This circumstance is explained by the fact that, among the substitutions of the λ, μ's, the following is found:

$$\lambda_1' = -\lambda_1,\ \lambda_2' = -\lambda_2,\ \mu_1' = -\mu_1,\ \mu_2' = -\mu_2,$$

for which the y_ν's remain altogether unaltered as bilinear functions of the λ, μ's.

We proceed to introduce the following substitution, which we describe shortly as an *interchange* of λ and μ:

$$(51) \qquad \mu_1' = \lambda_1,\ \mu_2' = \lambda_2,\ \lambda_1' = \mu_2,\ \lambda_2' = -\mu_1.$$

From the formula (40) we then get:

$$y_\nu' = y_{2\nu},$$

therefore, in fact, the uneven permutation of the y_ν's previously employed. In agreement likewise with what precedes, we get on repetition of (51):

$$\lambda_1' = \lambda_2,\ \lambda_2' = -\lambda_1;\ \mu_1' = \mu_2,\ \mu_2' = -\mu_1,$$

i.e., the homogeneous icosahedral substitution otherwise denoted by U.

Instead of the two-valued or symmetric homogeneous functions of the y_ν's, we shall now fix our attention altogether on such rational and, in particular, integral homogeneous functions (forms) of the λ_1, λ_2, as remain unaltered for the substitutions (49), (50), and (51) respectively. If this is only the case for (49) and (50), they are to be called *invariants* simply, while we will speak of *complete* invariants if invariance also occurs for (51). It may happen that an invariant merely changes its sign for (51); we then call it *alternating*. If an invariant is neither complete nor alternating, it will, in virtue of (51), be co-ordinated with a second. The relation of the two invariants is then mutual, for the repetition of (51) is an icosahedral substitution, and therefore leads back to the original invariant.

Evidently α, β, and γ are complete invariants, ∇ an alternating one. The forms which we have elsewhere used, $f(\lambda_1, \lambda_2)$, $H(\lambda_1, \lambda_2)$, $T(\lambda_1, \lambda_2)$, M_1, N_1, represent the more general type. Calling the first three f_1, H_1, T_1, for the sake of brevity, the forms which are derived by interchange of λ and μ shall be denoted by f_2, H_2, T_2, M_2, N_2.

§ 8. General Remarks on the Calculations which we have to Perform.

The statement of the question given in § 6 requires that certain *rational* invariants shall be expressed rationally in terms of α, β, γ, ∇. To this end we may first ask: what integral invariant functions (forms) are integral functions of α, β, γ, ∇? Evidently all those, and only those, which are integral functions of the y_ν's. But these are all such forms as *have the same degree in* λ_1, λ_2 *and* μ_1, μ_2 *respectively.* For, on the one hand, every integral function of the y_ν's certainly gives an integral function of the same degree in the λ's and the μ's, and, on the other hand, every form of the λ, μ's which is of the same degree in the λ's and μ's can be written in the form of an integral function of the terms $\lambda_1\mu_1$, $\lambda_2\mu_1$, $\lambda_1\mu_2$, $\lambda_2\mu_2$, and these terms are, disregarding numerical factors, equal to p_1, p_2, p_3, p_4, *i.e.*, integral functions of the y_ν's.*

On the basis of this theorem, our method will now be to so dispose a given *rational* invariant, which we are to represent as a rational function of α, β, γ, ∇, by the application of appropriate factors in the numerator and denominator, that the numerator and denominator, taken by themselves, are invariant forms of the same degree in the λ, μ's, and then to compute numerator and denominator individually as *integral* functions of α, β, γ, ∇.

Now, as regards the evaluation of such integral functions, we remark *that every invariant form of the same degree in* λ_1, λ_2 *and* μ_1, μ_2 *admits of being split up into a complete and an alternating invariant.* In fact, let F_1 be the proposed form, F_2 the co-ordinated form which arises from it by interchange of λ and μ. Let us then simply put:

$$F_1 = \frac{F_1 + F_2}{2} + \frac{F_1 - F_2}{2}. \tag{52}$$

* *Cf.* the analogous remark in the last paragraph of the preceding chapter.

Here $\frac{F_1 + F_2}{2}$, as a complete invariant, is an integral function of α, β, γ, alone, while $\frac{F_1 - F_2}{2}$, as an alternating invariant, breaks up into the product of ∇ and an integral function of α, β, γ.

The few rules thus established allow us to grapple with the computation of the magnitudes m_1, n_1, Z_1, by direct means.

§ 9. Fresh Calculation of the Magnitude m_1.

In our new notation:

$$m_1 = \frac{M_1}{12 f_1}. \tag{53}$$

We will now first multiply numerator and denominator by such an invariant form that there results on both sides the same degree in the λ, μ's. It is evidently simplest (though by no means necessary) to choose f_2 as such a factor. We thus write:

$$m_1 = \frac{M_1 f_2}{12 f_1 f_2}. \tag{54}$$

In this formula the denominator is in itself a complete invariant; but we subject the denominator to the splitting-up process just described. We thus obtain:

$$m_1 = \frac{(M_1 f_2 + M_2 f_1) + (M_1 f_2 - M_1 f_1)}{24 f_1 f_2}; \tag{55}$$

the computation of m_1 *is therefore reduced to replacing the two complete invariants:*

$$M_1 f_2 + M_2 f_1 \text{ and } f_1 f_2,$$

as well as the alternating invariant:

$$M_1 f_2 - M_2 f_1$$

by appropriate integral functions of α, β, γ *and of* α, β, γ, ∇ *respectively.*

We solve the problem which now lies before us by taking into consideration, on the one hand, the degree of the forms in the λ, μ's which have to be compared, and, on the other hand, returning to the explicit values of our forms in the λ, μ's (as we gave them in § 6). The invariants just mentioned $(M_1 f_2 + M_2 f_1)$,

&c., are respectively of degree 13, 12, and 13 in the λ, μ's. On the other hand, α, β, γ, ∇ exhibit, with respect to the same variables, the degrees 3, 4, 5, 10. Hence we conclude, in the first place, that $(M_1 f_2 + M_2 f_1)$ must be a linear combination of the terms $\alpha^3\beta$, $\alpha\gamma^2$, $\beta^2\gamma$, and then, further, that $f_1 f_2$ is equal to just such a combination of α^4, β^3, $\alpha\beta\gamma$; finally, that $(M_1 f_2 - M_2 f_1)$ coincides with $\alpha\nabla$, save as to a numerical factor. In order to compute the numerical coefficients still undetermined, it is sufficient to pay regard to a few terms only in the explicit values of the individual forms, say, then, to the leading terms which present themselves when we arrange the forms according to descending powers of λ_1 and ascending powers of λ_2. I communicate here, for the sake of completeness, the leading terms of the forms which we have to consider, each to the extent to which we shall actually use it. We find by § 6:

$$\begin{gathered} M_1 f_2 + M_2 f_1 \\ = \lambda_1^{13} (\mu_1^{12}\mu_2^{1} + 11\mu_1^{7}\mu_2^{6} - \mu_1^{2}\mu_2^{11}) + \lambda_1^{12}\lambda_2 (-26\mu_1^{10}\mu_2^{3} + 39\mu_1^{5}\mu_2^{8} \\ + \mu_2^{13}) + \ldots, \\ f_1 f_2 = \lambda_1^{11}\lambda_2 (\mu_1^{11}\mu_2 + 11\mu_1^{6}\mu_2^{6} - \mu_1\mu_2^{11}) + \ldots, \\ M_1 f_2 - M_2 f_1 = \lambda_1^{13}\mu_1^{12}\mu_2 + \ldots, \end{gathered}$$

likewise:

$$\begin{gathered} \alpha^3\beta = \lambda_1^{13}\mu_1^{7}\mu_2^{6} + \lambda_1^{12}\lambda_2 (-\mu_1^{10}\mu_2^{3} + 3\mu_1^{5}\mu_2^{8}) + \ldots, \\ \alpha\gamma^2 = \lambda_1^{13} (2\mu_1^{7}\mu_2^{6} + 2\mu_1^{2}\mu_2^{11}) + \lambda_1^{12}\mu_2 (0) + \ldots, \\ \beta^2\gamma = \lambda_1^{13} (\mu_1^{12}\mu_2 + 2\mu_1^{7}\mu_2^{6} + \mu_1^{2}\mu_2^{11}) + \lambda_1^{12}\lambda_2 (0) + \ldots, \\ \alpha^4 = \lambda_1^{12}\mu_1^{8}\mu_2^{4} + 4\lambda_1^{11}\lambda_2\mu_1^{6}\mu_2^{6} + \ldots, \\ \beta^3 = -\lambda_1^{12}\mu_1^{3}\mu_2^{9} + 3\lambda_1^{11}\lambda_2\mu_1^{6}\mu_2^{6} + \ldots, \\ \alpha\beta\gamma = -\lambda_1^{12} (\mu_1^{9}\mu_2^{3} + \mu_2^{3}\mu_2^{9}) + \lambda_1^{11}\lambda_2 (\mu_1^{11}\mu_2 + 10\mu_1^{6}\mu_2^{6} - \mu_1\mu_2^{11}) + \ldots, \\ \alpha\Delta = -\lambda_1^{13}\mu_1^{12}\mu_2 + \ldots \end{gathered}$$

From these values we now infer immediately:

$$(56) \qquad \begin{cases} M_1 f_2 + M_2 f_1 = 11\alpha^3\beta + 2\beta^2\gamma - \alpha\gamma^2, \\ f_1 f_2 = \alpha^4 - \beta^3 + \alpha\beta\gamma, \\ M_1 f_2 - M_2 f_1 = -\alpha\nabla, \end{cases}$$

and therefore finally:

$$(57) \qquad m_1 = \frac{(11\alpha^3\beta + 2\beta^2\gamma - \alpha\gamma^2) - \alpha\nabla}{24 (\alpha^4 - \beta^3 + \alpha\beta\gamma)},$$

which is exactly the value communicated in formula (38).

In the same manner we could now, of course, compute n_1

and Z_1 also: the calculations in question would only be somewhat more elaborate, because in them we are concerned with constructions of a higher degree in the λ, μ's. We shall be able to decompose these calculations, as we always can in similar cases, by proper principles of reduction into a greater number of smaller steps (compare Gordan's work). We do not enter further on this, because we have already, in § 5, obtained simple formulæ for n_1 and Z_1, and the principle of Gordan's method of computation will be sufficiently known by the example of m_1.

§ 10. Geometrical Interpretation of Gordan's Theory.

In the preceding paragraphs Gordan's theory has been expounded from a purely algebraical point of view: we shall bring it closer to our other considerations if we reflect briefly on its geometrical significance. We have here to interpret as *co-ordinates on the canonical surface* the ratios $\lambda_1 : \lambda_2$ and $\mu_1 : \mu_2$, as we regarded them in the last paragraph of the preceding chapter. An equation:

$$F(\lambda_1, \lambda_2; \mu_1, \mu_2) = 0$$

then defines a curve lying on the canonical surface, whose intersections with the generators of the first and second kinds are determined as regards number by the degree of F in μ and λ respectively. If F is an invariant, the curve in question is transformed into itself for the sixty even collineations, and is therefore *half regular* so far as it is irreducible. The curve becomes *regular*, on the same condition, if the invariant F is complete or alternating.

If we interpret in this sense the invariants occurring in the preceding paragraphs, we are merely led to curves whose significance is either immediately manifest or is *a priori* known. The curves $\alpha=0$, $\beta=0$, $\gamma=0$ have come before our notice above as curves of intersection of the canonical surface with the diagonal surface.*

* Employing for α the representation given in (41), we can now easily prove the assertion previously made, that the curve $\alpha=0$, *i.e.*, Bring's curve, possesses no true double points, is therefore irreducible, and belongs to the deficiency $p=4$.

$\nabla = 0$ gives a curve which evidently splits up into ten plane portions; $f_1 = 0$, $H_1 = 0$, $T_1 = 0$ represent certain aggregates of twelve, twenty, or thirty generators respectively of the first kind. But what do $M_1 = 0$, $N_1 = 0$ denote? It follows immediately from the form of M_1, N_1, that we have to do with curves of the fourteenth and eighth order respectively, which cut the individual generators of the first kind only once. *These are the same curves which we have before represented by formulæ of the following kind:*

(58)
$$\rho y_\nu = t_\nu(\lambda_1, \lambda_2) \,.\, W_\nu(\lambda_1, \lambda_2)$$
$$\rho \dot{y}_\nu = W_\nu(\lambda_1, \lambda_2).$$

In fact, we shall be led back to these formulæ if we determine from $M_1 = 0$, or $N_1 = 0$, the $\frac{\mu_1}{\mu_2}$ as a rational function of $\frac{\lambda_1}{\lambda_2}$, and insert the value found in the formulæ (40):

$$y_\nu = \epsilon^{4\nu}\lambda_1\mu_1 - \epsilon^{3\nu}\lambda_2\mu_1 + \epsilon^{2\nu}\lambda_1\mu_2 + \epsilon^{\nu}\lambda_2\mu_2.$$

In the same sense the equation:

(59)
$$m \,.\, T(\lambda_1, \lambda_2) \,.\, N_1 + 12n \,.\, f^2(\lambda_1, \lambda_2) \,.\, M_1 = 0$$

represents the whole family of those curves of the 38th order which we considered in § 4 of the present chapter (see formula (27)).

We now turn in particular to the computation of m_1 given in the preceding paragraph. Originally we had by (53):

$$m_1 = \frac{M_1}{12 f_1};$$

m_1 is therefore a function on the canonical surface, which vanishes along the curve of the 14th order, $M_1 = 0$, and becomes infinite for the twelve generators of the first kind $f_1 = 0$. Writing now, as was done in (54),

$$m_1 = \frac{M_1 f_2}{12 f_1 f_2},$$

we have evidently raised the two curves $M_1 = 0$, $f_1 = 0$, by addition of the curve $f_2 = 0$, *i.e.*, an aggregate of twelve generators of the second kind, to the *complete intersection* of the canonical surface with the accessory surface; the intersecting surfaces can then, in particular, be so chosen that they themselves are transformed into themselves for the 60 even collineations, and are therefore represented by equating to zero integral functions of α, β, γ, ∇. Hence the structure of formula (57), and also the measure of its arbitrariness might be made manifest. I leave it to the reader to interpret in a similar manner the significance of formulæ (36), (37), for Z_1 and n_1.

§ 11. Algebraical Aspects (after Gordan).

We have so far expounded the Gordan theory as it originated, viz., as a direct method for computing the magnitudes occurring in the solution of the canonical equations of the fifth degree. Herr Gordan has, however, in his exhaustive memoir published in the 13th volume of the Annalen, chosen a much higher standpoint; he has proposed to himself the problem: *to construct the full system of invariant forms* $F(\lambda_1, \lambda_2; \mu_1, \mu_2)$, *and as many relations as possible between these forms.* He thus finds 36 systems of forms, of which those which are different from α, β, γ, ∇ are connected by permutation of λ and μ. We cannot go more fully into these results, but must consider the method which Herr Gordan has employed for their deduction. Let us recall how we previously deduced $H(\lambda_1, \lambda_2)$, $T(\lambda_1, \lambda_2)$ from $f(\lambda_1, \lambda_2)$, by means of processes of differentiation appertaining to the invariant theory. In just the same way Herr Gordan obtains his forms, putting at the head of them:

$$\alpha = -\lambda_1^3\mu_1^2\mu_2 - \lambda_1^2\lambda_2\mu_2^3 - \lambda_1\lambda_2^2\mu_1^3 + \lambda_2^3\mu_1\mu_2^2$$

as a "double-binary ground-form with two series of independent variables."

Let us first explain, in respect to this, how $f(\lambda_1, \lambda_2)$ [the ground-form of the icosahedron] is now to be defined. Consider λ_1, λ_2 as constant in α, *i.e.*, α as a binary form of the third order of μ_1, μ_2 only. *Then,* I assert, *f is the discriminant of this form of the third order, disregarding a numerical factor.* We confirm this by direct calculation. We first construct, in accordance with the usual rules, the Hessian form of α, and find, save as to a factor, the following invariant, quadratic in the μ's:

(60) $\tau = \mu_1^2(-\lambda_1^6 - 3\lambda_1\lambda_2^5) + 10\mu_1\mu_2\lambda_1^3\lambda_2^3 + \mu_2^2(3\lambda_1^5\lambda_2 - \lambda_2^6)$,

which we shall again use later on. We further compute the determinant of τ, and come back, in fact, disregarding a numerical coefficient, to:

$$f = \lambda_1^{11}\lambda_2 + 11\lambda_1^6\lambda_2^6 - \lambda_1\lambda_2^{11}.$$

Let us further explain how Herr Gordan obtains the formulæ of inversion, which we were able to establish in § 4 by applying those data which we obtained above, I, 4, § 12 (somewhat incidentally), by the formation of the canonical resolvent of the icosahedron equation. In Herr Gordan's method those invariants which are *linear* in μ_1, μ_2 form the starting-point. He shows that four different invariants of this kind exist, among which those two which are of lowest degree in the λ's are exactly identical * with our N_1, M_1. Now by formula (40) the y_ν's are themselves linear forms in μ_1, μ_2:

$$y_\nu = \epsilon^{4\nu}\lambda_1\mu_1 - \epsilon^3\lambda_2\mu_1 + \epsilon^{2\nu}\lambda_1\mu_2 + \epsilon^\nu\lambda_2\mu_2.$$

Hence we can write, from the outset:

(61) $$ay_\nu = b_\nu \,.\, M_1 + c_\nu \,.\, N_1,$$

where the coefficients a, b_ν, c_ν are to be taken from the identity:

$$\begin{vmatrix} y_\nu & M_1 & N_1 \\ \frac{\delta y_\nu}{\delta\mu_1} & \frac{\delta M_1}{\delta\mu_1} & \frac{\delta N_1}{\delta\mu_1} \\ \frac{\delta y_\nu}{\delta\mu_2} & \frac{\delta M_1}{\delta\mu_2} & \frac{\delta N_1}{\delta\mu_2} \end{vmatrix} = 0.$$

Here a, as the functional determinant of M_1 and N_1, is itself an invariant; we have seen above that it is identical with $H(\lambda_1, \lambda_2)$. On the other hand, b_ν, c_ν are necessarily five-valued, like the y_ν's themselves. Computing them as functional determinants of y_ν and N_1, y_ν and M_1 respectively, we then get the same magnitudes as we before denoted by $W_\nu(\lambda_1, \lambda_2)$ and $t_\nu(\lambda_1, \lambda_2) \,.\, W_\nu(\lambda_1, \lambda_2)$. In fact, formula (61), written in our earlier notation, must run as follows:

* One of these four invariants, if we multiply it by H_1, is contained in the general form $m \,.\, T_1 \,.\, N_1 + 12n \,.\, f_1^2 \,.\, M_1$, the vanishing of which represents those curves of the 38th order which we have previously considered. Among these curves, in addition to $M_1 = 0$, $N_1 = 0$, a third presents itself whose order reduces to a lower number, viz., to 18.

(62) $H(\lambda_1, \lambda_2) \cdot y_\nu = W_\nu(\lambda_1, \lambda_2) \cdot M_1 + t_\nu(\lambda_1, \lambda_2) \cdot W_\nu(\lambda_1, \lambda_2) \cdot N_1$;

compare, say, formula (46) *supra*. We can say that Gordan's development of this formula, as we have explained it, is just the reverse of ours. The further course of the calculation is then the same in both. In order to express the y_ν's by means of the λ's and the other given magnitudes, we introduce in (62), instead of M_1, N_1, the expressions:

$$m_1 = \frac{M_1}{12 f_1}, \quad n_1 = \frac{N_1 \cdot T_1}{144 \cdot f_1^3},$$

i.e., quotients which are both of the first dimension in λ_1, λ_2 and μ_1, μ_2, and then compute these as rational functions of α, β, γ, ∇, in the same way as was done in § 9 for m_1 in particular.

We pause for another moment over Gordan's derivation of formula (62). We can evidently put it into words as follows. Since the y_ν's are bilinear forms in λ_1, λ_2 and μ_1, μ_2, their determination requires (if we have assumed λ_2 arbitrarily—as is allowed—and then found $\lambda_1 : \lambda_2$ from the corresponding icosahedral equation) only the knowledge of μ_1, μ_2 in addition. *We now obtain these by annexing the two invariants, linear in* μ_1, μ_2, *to wit*, M_1, N_1, *and compute them as rational functions of* λ_1, λ_2, *and of* α, β, γ, ∇. In fact, we have thus two linear equations for μ_1, μ_2; if we solve these for μ_1, μ_2, and insert the values which arise in the formula for y_ν, we have the result which we sought, the same which is presented in an abbreviated form by (62). Or we can also put it thus. If we put $M_1 = 0$, we determine in the binary manifoldness $\mu_1 : \mu_2$ a first element *contragredient* to the elements $\lambda_1 : \lambda_2$, or—to express it more generally—a *covariant* element. We obtain a second element of the same kind if we take $N_1 = 0$. Our problem is to find that element in the manifoldness $\mu_1 : \mu_2$ which is represented by $y_\nu = 0$. We solve this problem in (62) by constructing y_ν with the two covariant elements M_1 and N_1 by the help of appropriate coefficients, thus proceeding according to the same fundamental theorems of the "typical representation" which we employed above in describing the Tschirnhausian transformation. The mode of conception thus denoted will often come into play later on in a generalised form.

§ 12. The Normal Equation of the r_ν's.

In our general survey of the different paths struck out for solving the equation of the fifth degree, we have about (II, 1, § 1) separated the method of *resolvent construction* from that of the *Tschirnhausian transformation*, remarking, however, that we can always replace the one method by the other. When we solved the canonical equations of the fifth degree directly by the help of the icosahedral equation, we followed the method of resolvent construction. If we are to expound the method of the Tschirnhausian transformation in place of it, we shall have to start from one of the resolvents of the fifth degree, which we have established in I, 4, for the icosahedral equation, as a *normal equation.*

The resolvent of the r_ν's which we constructed in § 9, *loc. cit.*, and to which we then assigned the form:

$$(63) \qquad Z : Z-1 : 1 = (r-3)^3(r^2-11r+64) : r(r^2-10r+45)^2 : -1728,$$

seems to be in this respect most adapted to our purpose. In fact, we have already (*supra*, § 13, *loc. cit.*) represented the u_ν, v_ν, rationally in terms of r_ν:

$$u_\nu = \frac{12}{r_\nu^2 - 10r_\nu + 45}, \quad v = \frac{12}{r_\nu - 3};$$

if we insert these formulæ in our present one:

$$y = m_1 \cdot u_\nu + n_1 \cdot u_\nu \cdot v_\nu$$

we obtain immediately the representation of y_ν by means of the roots of the normal equation (63):

$$(64) \qquad y = \frac{12(r_\nu - 3)m_1 + 144n_1}{(r_\nu - 3)(r_\nu^2 - 10r_\nu + 45)}.$$

The only further question that arises is how we are to compute

$$r = \frac{t_\nu^2(\lambda_1, \lambda_2)}{f(\lambda_1, \lambda_2)}$$

as a rational function of the y_ν's. We will here strike out a path similar to that just taken (§ 9) in the computation of m_1. For brevity, let:

$$t_\nu(\lambda_1, \lambda_2) = t_{\nu,\,1}, \quad t_\nu(\mu_1, \mu_2) = t_{\nu,\,2},$$

then we write in turn:

$$r_\nu = \frac{t_{\nu,\,1}^2 \cdot f_2}{f_1 \cdot f_2},$$
$$= \frac{[t_{\nu,\,1}^2 \cdot f_2 + t_{\nu,\,2}^2 \cdot f_1] + [t_{\nu,\,1}^2 \cdot f_2 - t^2_{,\,2} \cdot f_1]}{2 f_1 \cdot f_2}.$$

Here $f_1 \cdot f_2$ is, as we know, equal to $(\alpha^4 - \beta^3 + \alpha\beta\gamma)$. We can now compute, in a perfectly analogous manner, the two portions of the numerator (by returning to the explicit values in λ_1, λ_2 and μ_1, μ_2). Let us for a moment suppose the y's to be introduced in place of the λ, μ's, then these components are such integral functions of the y's as remain unaltered *when we permute those four y's, which are different from our fixed y_ν, in an arbitrary manner and in an even manner respectively.* Now the sums of the powers of these four y's are integral functions of y_ν, α, β, γ, but their difference-product is equal to $5\nabla : (y_\nu^4 + 2\alpha y_\nu + \beta)$, where $(y^4 + 2\alpha y + \beta)$ denotes the differential coefficient of the left side of our principal equation divided by 5. *Hence $[t_{\nu,\,1}^2 \cdot f_2 + t_{\nu,\,2}^2 \cdot f_1]$ will be an integral function of y_ν, α, β, γ, but $[t^2_{,\,1} \cdot f_2 - t_{\nu,\,2}^2 \cdot f_2]$ will split up into the product of such an integral function and the magnitude* $\frac{\nabla}{y_\nu^4 + 2\alpha y_\nu + \beta}$. It is not necessary for me to go into the details of the calculation; I will therefore only communicate the result.* We find:

(65)
$$2(\alpha^4 - \beta^3 + \alpha\beta\gamma)\, r_\nu$$
$$= [(\alpha\gamma + 2\beta^2)\, y_\nu^4 + (\alpha^3 - \beta\gamma)\, y_\nu^3 - 5\alpha^2\beta \,.\, y_\nu + (4\alpha^2\gamma + 13\alpha\beta^2)\, y_\nu$$
$$+ (11\alpha^4 + 9\alpha\beta\gamma)] - \left[(\alpha y_\nu^3 + \beta y_\nu^2 + \alpha^2) \cdot \frac{\nabla}{y_\nu^4 + 2\alpha y_\nu + \beta} \right].$$

Summing up, we have the following result: *We have in* (65) *the Tschirnhausian transformation which transforms the given canonical equation into the normal equation* (63); *if we have then determined the roots r_ν of the latter,* (64) *gives us the explicit values of the y_ν's which we sought.*

* See Math. Ann., t. xii, p. 556.

§ 13. Bring's Transformation.

I have communicated in detail the formulæ of the preceding paragraph the more willingly because from them, as I shall now show, all formulæ can be derived which are required in the execution of the *Bring transformation.** Let y_0, y_1, y_2, y_3, y_4 and $y_0', y_1', y_2', y_3', y_4'$ be the co-ordinates of two points on the canonical surface which belong to the same generator of the first kind. Then we obtain for the corresponding canonical equation's the Z and r_ν's,† while we distinguish the other magnitudes which present themselves therein for consideration by addition of an accent, and will therefore put $\alpha', \beta', \gamma', \nabla'$, m_1', n_1' in the second equation opposite $\alpha, \beta, \gamma, \nabla, m_1, n_1$ in the first. I say now that a double application of the formulæ (64), (65) is sufficient in order to transform one of the canonical equations into the other, and the roots of the one into those of the second. We will, for brevity, denote by (64′), (65′) the equations (64), (65) when they are written with accentuated letters. Then the whole process which is here necessary evidently consists in expressing first, by means of (65) the r_ν's in terms of the y_ν's, and then, by means of (64′), the y_ν''s in terms of the r_ν's (which is the transformation we sought), and then, conversely, computing by means of (65′) the r_ν's as functions of the y_ν's, and so finding from them the y_ν's by means of (64).

The Bring theory is furnished by a special case of the general method thus given. The generator of the first kind, viz., which carries the point y, meets the curve $\alpha = 0$ in three points: we obtain the Bring transformation if we choose one of these points as y'. This means, analytically, that we are so to determine m_1', n'_1 that, in the canonical equation for y', the term involving y^2 disappears. A glance at the general canonical resolvent (I, 4, § 12) gives us at once the cubic equations which m_1', n_1' must satisfy in consequence; in other words, the cubic auxiliary equation which the Bring theory required; it is as follows:

* See the analogous formulæ in *Gordan's* paper in Bd. xiii of the Math. Ann., p. 400, &c.

† Or more correctly Z_1 and $r_{\nu,1}$'s, as we might have written in the preceding paragraph.

(66) $$8m^3+12m^2n+\frac{6mn^2+n^3}{1-Z}=0.$$

It depends, as is clear *a priori*, not on the individual point y, but only on the generator of the first kind on which this point is situated, and on the sixty generators which arise from the one mentioned in virtue of the even collineations. We have nothing further to add concerning the Bring theory; the most we can do is to call attention to the fact that (65′) now becomes very simple, inasmuch as $\alpha'=0$.* It will also be useful to give prominence to the fact that, in the trinomial equation which we obtain by carrying out the Bring transformation, we always know, *a priori*, the square root of the discriminant.

§ 14. The Normal Equation of Hermite.

Now that we have brought the Bring theory so simply into connection with our developments, we will seek to do the same with the normal form on which *Hermite* bases the solution by elliptic functions. As we saw above (II, 1, § 4), this runs as follows:

(67) $$Y^5-2^4 \cdot 5^3 \cdot u^4(1-u^8)^2 \cdot Y-2^6\sqrt{5^5} \cdot u^3(1-u^8)^2(1+u^8)=0,$$

where $u^8=\kappa^2$. We shall inquire if this equation is contained as a special case in the general canonical resolvent of the icosahedral equation, when we put Z, the right side of the icosahedral equation, equal to:

(68) $$\frac{g_2{}^3}{\Delta}=\frac{4}{27}\cdot\frac{(1-\kappa^2+\kappa^4)^3}{\kappa^4(1-\kappa^2)^2},$$

as we did above (I, 5, § 7) in dealing with the solution of the icosahedral equation by elliptic modular functions; we shall inquire why Hermite in his investigations was led, at the outset, to the Bring form, while every canonical equation of

* In a manner similar to that in which the Bring transformation is effected by means of (66), the problem is solved by the help of an equation of the fourth degree: from the given canonical equation to establish another for which $\beta'=0$. To the feasibility of this problem it seems that *Jerrard* first called attention [Mathematical Researches, 1834].

the fifth degree can be solved by the help of elliptic functions (through the intervention of the icosahedral equation), and the Bring form is by no means the simplest among the infinitely many canonical equations with one parameter which present themselves.

In order to answer these questions, let us insert for Z in (66) the function of κ^2 given in (68). *The result is that the cubic equation* (66) *becomes reducible.* In fact, it is satisfied, as we can verify immediately, if we choose:

$$m : n = 3\kappa^2 : 2\,(2 - 5\kappa^2 + 2\kappa^4).$$

I will accordingly put:

$$(69) \qquad m = 3\kappa^2\,(1 + \kappa^2), \quad n = 2\,(1 + \kappa^2)\,(2 - 5\kappa^2 + 2\kappa^4).$$

The coefficients of the canonical resolvent given in I, 4, § 12, are then considerably condensed, so that we obtain the equation:

$$(70) \quad y^5 - 2^4 \,.\, 3^8 \,.\, 5 \,.\, \kappa^{10}\,(1 - \kappa^2)^2 \,.\, y - 2^6 \,.\, 3^{10} \,.\, \kappa^{12}\,(1 - \kappa^2)^2\,(1 + \kappa^2) = 0.$$

Here we need only further substitute for y:

$$(71) \qquad y = \frac{\sqrt{5}}{9\kappa^{\frac{9}{4}}} \,.\, Y,$$

in order to find precisely the Hermite equation.

Our first question is therefore to be answered in the affirmative. At the same time we discern the answer to the second question in the circumstance that Hermite operated, not with the rational invariants g_2, g_3, *but with* κ^2 *throughout.*

If we now compute for the Hermitian equation, or, what comes to the same thing, for (70), the corresponding Z_1, we naturally come back, a proper choice of the sign of ∇ being made, to $\frac{g_2^3}{\Delta}$.* But a very simple value arises for Z_2 also; we find, on reversing the sign of ∇ in the expression for Z_1:

$$(72) \qquad Z_2 = \frac{(1 + 14\kappa^2 + \kappa^4)^3}{108\kappa^2\,(1 - \kappa^2)^4}.\dagger$$

* We have here to take (for (70)):

$$\nabla = 2^{12} \,.\, 3^{20} \,.\, \kappa^{24}\,(1 - \kappa^2)^4\,(1 - 6\kappa^2 + \kappa^4).$$

† *Cf. Gordan, loc. cit.*, or my communication, already mentioned, in the Rendiconti of the Istituto Lombardo of April 26, 1877.

This, as is shown in the theory of elliptic functions, is one of the three values which arise from $\frac{g_2^3}{\Delta}$ *by a quadratic transformation of the elliptic integral.* We cannot, unfortunately, follow up further in this place the interesting connection of the Bring curve with the quadratic transformation of elliptic functions which here presents itself.*

Breaking off for the present these developments, we here content ourselves with the fact that the Bring and Hermitian formulæ fit in with ours. In the fifth chapter we shall return to our present results from a general point of view, and seek to decide what theoretical value they possess.

* *Cf.* my memoir: "Ueber die Transformation der elliptischen Functionen und die Auflösung der Gleichungen fünften Grades," in Bd. xiv of the Math. Annalen (1878), especially p. 166, &c., of the same.

CHAPTER IV

THE PROBLEM OF THE A's AND THE JACOBIAN EQUATIONS OF THE SIXTH DEGREE

§ 1. The Object of the Following Developments.

In the preceding chapter we have considered two series of binary variables, λ_1, λ_2 and μ_1, μ_2, which were simultaneously subjected to homogeneous icosahedral substitutions, and besides to a process which we called the interchange of λ, μ. We have further had under investigation certain bilinear forms of the λ, μ's which we called y_ν. The y_ν's undergo on their part, for the transformations of the λ, μ's in question, linear substitutions of the simplest possible kind, to wit, mere permutations, and indeed permutations of the whole set; if we are therefore to establish a corresponding *form-problem of the y's*, this finds its complete expression in the equation of the fifth degree which the y_ν's satisfy, *i.e.*, in the *canonical equation.* We can in this sense assert that we have been concerned in the preceding chapter with a form-problem which arises from the consideration of the simultaneous substitutions of the λ, μ's.

Now in the following pages a statement of the question of a quite similar kind (which moreover possesses essentially a still more simple character) is to be dealt with. The simultaneous icosahedral substitutions of the λ, μ's were, as we called it, contragredient; *we will now take into consideration two series of binary variables:*

$$\lambda_1, \lambda_2;\ \lambda_1', \lambda_2',$$

which are in each case simultaneously subjected to the same icosahedral substitutions, and so can be described as cogredient. In the case of these, again, we construct certain bilinear forms, viz., the symmetric functions:

(1) $$A_0 = -\frac{1}{2}(\lambda_1\lambda_2' + \lambda_2\lambda_1'), \quad A_1 = \lambda_2\lambda_2', \quad A_2 = -\lambda_1\lambda_1',$$

i.e., the coefficients of that quadratic form:

(2) $$A_1 z_1^2 + 2A_0 z_1 z_2 - A_2 z_2^2,$$

which arises on multiplying out the factors:

$$\lambda_2 z_1 - \lambda_1 z_2, \quad \lambda_2' z_1 - \lambda_1' z_2.$$

If we subject the λ, λ''s to the 120 homogeneous icosahedral substitutions, or interchange them with one another, these **A**'s undergo on the whole sixty ternary linear substitutions, for the individual **A**'s remain altogether unaltered, not only for the interchange of the λ, λ''s, but also when we simultaneously reverse the signs of λ_1, λ_2, λ_1', λ_2'.* *We shall deal with the ternary form-problem which is involved in the consideration of the substitutions thus defined.*

We have already stated that this form-problem of the **A**'s is essentially more simple than that of the y's. In fact, we shall be able to direct our considerations and computations throughout towards the ordinary icosahedral problem, from which the results we seek then offer themselves in virtue of a definite *principle of transference* well known in modern algebra, so that indeed the accomplishment of our problem appears almost as an exercise in the application of certain fundamental theorems appertaining to the theory of invariants.† According to the same scheme, we should also be able to deal with the case of 3, 4 . . . series of binary variables which are subjected to the icosahedral substitutions or any other group of binary substitutions in a cogredient manner. If among these infinitely many, so to say, associated form-problems we select the one just described, it is because we employ it in the further consideration of equations of the fifth degree. We shall soon learn *that the general Jacobian equations of the sixth degree, by which Kronecker's theory of equations of the fifth degree is supported, are resolvents of our problem of the* **A**'s. By substituting for it

* The substitutions of the A's are hence simply isomorphic with the sixty ordinary non-homogeneous icosahedral substitutions.

† The principle of transference in question is essentially the same as that to which Hesse has devoted a memoir in Bd. lxvi of Crelle's Journal (1866).

altogether the problem of the **A**'s, we shall succeed in the simplest way in understanding from our standpoint the various results which have been discovered in other quarters for the Jacobian equations of the sixth degree, and thus in attaining for the general treatment of equations of the fifth degree a uniform basis, which is nothing else than a rational theory of the icosahedron.*

The arrangement of the subject-matter for the following developments is already given by what we have said. The first thing is to establish the problem of the **A**'s in an explicit form, where we shall again make free use of geometrical interpretation. On then studying the corresponding resolvents, we are enabled to pass over to the Jacobian equations of the sixth degree, and to the researches of Brioschi and Kronecker relative thereto. I finally apply myself to the solution of our problem, and show that it can be accomplished with the help of an icosahedral equation and an additional square root, in strict analogy with the Gordan theory expounded in the preceding chapter.†

§ 2. The Substitutions of the A's—Invariant Forms.

In order now to determine explicitly the substitutions of our **A**'s, let us recur to the generating icosahedral substitutions S, T, and U respectively. We had for the λ_1, λ_2's:

$$(3)\qquad \begin{cases} S: & \lambda_1' = \pm\epsilon^3\lambda_1, \quad \lambda_2' = \pm\epsilon^2\lambda_2; \\ T: & \begin{cases} \sqrt{5}\,.\,\lambda_1' = \mp(\epsilon-\epsilon^4)\lambda_1 \pm (\epsilon^2-\epsilon^3)\lambda_2, \\ \sqrt{5}\,.\,\lambda_2' = \pm(\epsilon^2-\epsilon^3)\lambda_1 \pm(\epsilon-\epsilon^4)\lambda_2; \end{cases} \\ U: & \lambda_1' = \mp\lambda_2, \quad \lambda_2' = \pm\lambda_1. \end{cases}$$

* Like the Jacobian equations of the sixth degree, the general equations of the $(n+1)^{\text{th}}$ degree which we described above (II, 1, § 13) can be replaced by parallel form-problems which are related to $\frac{(n+1)}{2}$ variables A_0, A_1, . . . $A_{\frac{n-1}{2}}$. I have accomplished this for $n=7$ in Bd. xv of the Math. Ann. (1879); see in particular pp. 268–275.

† The principal reflexions to be employed in the following expositions were laid before the Erlanger Societät by me on November 18, 1876 ["Weitere Untersuchungen über das Ikosader, I"]; *cf.* further the second part of my memoir, which appeared under the same title in Bd. xvii of the Annalen (1877). The developments § 8–13 were then added for the first time.

Writing* down the same formulæ for the λ'_1, λ'_2, we obtain from (1) for our **A**'s the following substitutions:

$$(4)\quad \begin{cases} S:\ \mathsf{A}_0' = \mathsf{A}_0,\quad \mathsf{A}_1' = \epsilon^4\mathsf{A},\quad \mathsf{A}_2' = \epsilon\mathsf{A}_2\,; \\ T: \begin{cases} \sqrt{5}\,.\,\mathsf{A}_0' = \mathsf{A}_0 + \mathsf{A}_1 + \mathsf{A}_2, \\ \sqrt{5}\,.\,\mathsf{A}_1' = 2\mathsf{A}_0 + (\epsilon^2+\epsilon^3)\mathsf{A}_1 + (\epsilon+\epsilon^4)\mathsf{A}_2, \\ \sqrt{5}\,.\,\mathsf{A}_2' = 2\mathsf{A}_0 + (\epsilon+\epsilon^4)\mathsf{A}_1 + (\epsilon^2+\epsilon^3)\mathsf{A}_2\,; \end{cases} \\ U:\ \mathsf{A}_0' = -\mathsf{A}_0,\quad \mathsf{A}_1' = -\mathsf{A}_2,\quad \mathsf{A}_2' = -\mathsf{A}_1, \end{cases}$$

which all, like (3), have the determinant $+1$. From them are composed the 60 linear substitutions of the **A**'s which exist according to the old scheme (I, 1, § 12):

$$(5)\qquad S^\mu,\quad S^\mu T S^\nu,\quad S^\mu U,\quad S^\mu T S^\nu U\ (\mu,\ \nu = 0,\ 1,\ 2,\ 3,\ 4).$$

Now, as regards the invariant forms, *i.e.*, those integral homogeneous functions of the **A**'s which remain unaltered for the substitutions (5), *the determinant of* (2):

$$(6)\qquad A = \mathsf{A}_0^2 + \mathsf{A}_1\mathsf{A}_2,$$

at all events, belongs to them. In fact, this becomes, on introducing the λ, λ''s equal to $(\lambda_1, \lambda_2' - \lambda_2, \lambda_1')^2$, and therefore remains altogether invariant if we subject the λ, λ''s simultaneously to any homogeneous substitutions of determinant 1. *Besides A, the full system of the forms which we seek will only contain, as I assert, three more forms, of the* 6*th*, 10*th*, *and* 15*th degrees respectively*. Namely, if $A = 0$, then $\lambda_1' = M\lambda_1$, $\lambda_2' = M\lambda_2$, understanding by M an arbitrary number; therefore by (1):

$$(7)\qquad \mathsf{A}_0 = -M\lambda_1\lambda_2,\quad {}_1 = M\lambda_2^2,\quad \mathsf{A}_2 = -M\lambda_1^2.$$

The required forms are accordingly transformed into multiples of forms of λ_1, λ_2, whose degree in the λ's is double as great as the original degree in the **A**'s, and which, moreover, have the property of being transformed into themselves by the homogeneous icosahedral substitutions of λ_1, λ_2. But now the system of all icosahedral forms is composed of the form of the 12th order $f(\lambda_1, \lambda_2)$, the form of the 20th order $H(\lambda_1\ \lambda_2)$, and the form of the 30th $T(\lambda_1, \lambda_2)$. Hence follows our assertion by

* I hope no misunderstanding will arise from the fact that the letters λ_1', λ_2' just employed in formula (3) on the left-hand side have been used before, of course with quite another meaning.

reciprocation. We might even say that, corresponding to the identity:

$$(8) \qquad T^2 = 1728f^5 - H^3,$$

a single identical relation will subsist between the new forms which is transformed into (8) as soon as we put $A=0$.

I will denote the three required forms by B, C, D. In proving their existence by reverting to the icosahedral forms f, H, T, we have already made use of the algebraical principle of transference which we proposed above. This we shall do in a higher measure by now actually establishing B, C, D, if only in a provisional form. We are here dealing with a process of *polarisation* adapted to the purpose. If $\phi(\lambda_1, \lambda_2)$ is any form which remains unaltered for the homogeneous icosahedral substitutions of λ_1, λ_2, and if λ_1', λ_2', are cogredient with λ_1, λ_2, then all the polars:

$$\frac{\delta\phi}{\delta\lambda_1}\cdot\lambda_1' + \frac{\delta\phi}{\delta\lambda_2}\cdot\lambda_2',$$

$$\frac{\delta^2\phi}{\delta\lambda_1^2}\cdot\lambda_1'^2 + 2\frac{\delta^2\phi}{\delta\lambda_1\delta\lambda_2}\cdot\lambda_1'\lambda_2' + \frac{\delta^2\phi}{\delta\lambda_2^2}\cdot\lambda_2'^2, \text{ \&c.},$$

will be invariant for the simultaneous substitutions of the λ, λ''s. Let us now construct in particular for $f_1(\lambda_1, \lambda_2)$, $H(\lambda_1, \lambda_2)$, $T(\lambda_1, \lambda_2)$, respectively the *sixth*, *tenth*, and *fifteenth* polars. We thus obtain invariant forms which are symmetrical in the λ, λ''s, and therefore represent integral functions of $\mathsf{A}_0, \mathsf{A}_1, \mathsf{A}_2$. On writing them down as such, we have found the required forms B, C, D. In fact, these forms are now necessarily invariant for the substitutions (4) or (5); they have, moreover, the degrees 6, 10, 15 in the A's, and are transformed into multiples of $f(\lambda_1, \lambda_2)$, $H(\lambda_1, \lambda_2)$, $T(\lambda_1, \lambda_2)$, when the formulæ (7) are applied. I will communicate here at once the result of the calculation. After separating particular numerical factors, we find in the manner explained:

$$(9)\left\{\begin{aligned}
B' &= 16\mathsf{A}_0^6 - 120\mathsf{A}_0^4\mathsf{A}_1\mathsf{A}_2 + 90\mathsf{A}_0^2\mathsf{A}_1^2\mathsf{A}_2^2 + 21\mathsf{A}_0(\mathsf{A}_1^5 + \mathsf{A}_2^5) - 5\mathsf{A}_1^3\mathsf{A}_2^3,\\
C' &= -512\mathsf{A}_0^{10} + 11520\mathsf{A}_0^8\mathsf{A}_1\mathsf{A}_2 - 40320\mathsf{A}_0^6\mathsf{A}_1^2\mathsf{A}_2^2 + 33600\mathsf{A}_0^4\mathsf{A}_1^3\mathsf{A}_2^3\\
&\quad - 6300\mathsf{A}_0^2\mathsf{A}_1^4\mathsf{A}_2^4 - 187(\mathsf{A}_1^{10} + \mathsf{A}_2^{10}) + 126\mathsf{A}_1^5\mathsf{A}_2^5\\
&\quad + \mathsf{A}_0(\mathsf{A}_1^5 + \mathsf{A}_2^5)(22176\mathsf{A}_0^4 - 18480\mathsf{A}_0^2\mathsf{A}_1\mathsf{A}_2 + 1980\mathsf{A}_1^2\mathsf{A}_2^2),\\
D &= [\mathsf{A}_1^5 - \mathsf{A}_2^5]\{-1024\mathsf{A}_0^{10} + 3840\mathsf{A}_0^8\mathsf{A}_1\mathsf{A}_2 - 3840\mathsf{A}_0^6\mathsf{A}_1^2\mathsf{A}_2^2\\
&\quad + 1200\mathsf{A}_0^4\mathsf{A}_1^3\mathsf{A}_2^3 - 100\mathsf{A}_0^2\mathsf{A}_1^4\mathsf{A}_2^4 + \mathsf{A}_1^{10} + \mathsf{A}_2^{10} + 2\mathsf{A}_1^5\mathsf{A}_2^5\\
&\quad + \mathsf{A}_0(\mathsf{A}_1^5 + \mathsf{A}_2^5)(352\mathsf{A}_0^4 - 160\mathsf{A}_0^2\mathsf{A}_1\mathsf{A}_2 + 10\mathsf{A}_1^2\mathsf{A}_2^2)\}.
\end{aligned}\right.$$

I have here denoted the two first forms no longer by B and C, but by B' and C', because I will hereafter modify these further by addition of factors which contain A as factor. Only when this is done shall I establish the relation which makes D^2 equal to an integral function of A, B, C. If we apply the substitution (7) to the preceding forms, which for simplicity we will put $M=1$, we have, in agreement with what was said before:

$$(10) \qquad \begin{cases} B' = 21 \cdot f(\lambda_1, \lambda_2), \\ C' = 187 \cdot H(\lambda_1, \lambda_2), \\ D = \quad \cdot T(\lambda_1, \lambda_2).^* \end{cases}$$

§ 3. Geometrical Interpretation—Regulation of the Invariant Expressions.

In order to facilitate our mode of expression and the growth of our ideas in the domain of the function-theory, let us now introduce our geometrical interpretation. Retaining throughout the analogy with the developments of the preceding chapter, let us regard $\mathsf{A}_0 : \mathsf{A}_1 : \mathsf{A}_2$ as the projective co-ordinates of a point in the plane, the substitutions of the A's as so many plane collineations.† The individual invariant form of the A's

* The method of calculation contained in the text is described in the text-books on the theory of invariants, at the suggestion of *Gordan*, as *transvection* of the quadratic forms (2), and indeed (disregarding numerical factors) B' is the 6th, C' the 10th, D' the 15th, transvectant of the corresponding power of (2) over f, H, and T respectively. I have not applied this mode of expression and the corresponding symbolical relation in the text, because I wished not to presuppose in this respect any specific preliminary knowledge on the part of the reader.

† We can of course regard every form-problem in a corresponding manner. If we proceeded differently in the foregoing part, and interpreted the binary form-problem by means of points of the $(x+iy)$ sphere, it was because we wished to have *intuitively* before our eyes not only the real, but also the complex values of the variables in the elementary sense.

I annex hereto a somewhat different interpretation of the problem of the A's. Put $\mathsf{A}_0=z$, $\mathsf{A}_1=x+iy$, $\mathsf{A}_2=x-iy$, and regard x, y, z as the rectangular co-ordinates of a point in space. Observing that the sixty substitutions of the A's have the determinant unity, and A is now $=x^2+y^2+z^2$, we recognise that the said substitutions now correspond to *rotations round the origin of co-ordinates*. These are the rotations for which a determinate icosahedron is brought into coincidence with itself. The six fundamental points to be immediately introduced in the text give on this interpretation those six diameters which connect two opposite summits of the icosahedron. On the then represents, when equated to zero, a plane curve which is transformed into itself for the said collineations. In this respect we have first the conic $A=0$, which we will call the *fundamental conic.* If we write in accordance with formulæ (7) (again taking $M=1$):

$$\mathsf{A}_0 = -\lambda_1\lambda_2, \quad \mathsf{A}_1 = \lambda_2^2, \quad \mathsf{A}_2 = -\lambda_1^2,$$

we have expressed the variable point of this conic by means of a parameter $\frac{\lambda_1}{\lambda_2}$. Hence we shall be able to denote the two parameters $\frac{\lambda_1}{\lambda_2}, \frac{\lambda'_1}{\lambda'_2}$, which appear in formulæ (1) by means of two points of the fundamental conic. *These are the two points in which the two tangents from the point* **A** *to the fundamental conic touch the latter.* In fact, the polar of the point **A** with respect to $A=0$ has the equation:

$$2\mathsf{A}_0\mathsf{A}_0' + \mathsf{A}_2\mathsf{A}_1' + \mathsf{A}_1\mathsf{A}_2' = 0,$$

and this equation is satisfied if we substitute for the **A**'s the expressions (1), and for the **A**''s the expressions (7) or the corresponding ones in which λ' is written instead of λ.

The points of the fundamental conic are naturally so grouped that aggregates of twelve, twenty, thirty of them are self-conjugate, represented respectively by:

$$f(\lambda_1, \lambda_2) = 0, \quad H(\lambda_1, \lambda_2) = 0, \quad T(\lambda_1, \lambda_2) = 0;$$

these are at the same time the points of intersection of $A=0$ with the curves $B=0$, $C=0$, $D=0$. We will now connect by a straight line those pairs amongst these points which remain fixed for the same collineations. Then we obtain, corresponding to the forms f, H, T, six, ten, and fifteen straight lines respectively. Constructing, then, for each of these lines, its pole with respect to the fundamental conic, we obtain self-conjugate groups of six, ten, and fifteen points in the plane.

Let us now consider the form of the equation:

$$A = \mathsf{A}_0^2 + \mathsf{A}_1\mathsf{A}_2 = 0.$$

other hand, the equation $D=0$ (of which we shall presently show that it splits up into fifteen linear factors) gives the fifteen planes of symmetry of the configuration.

We can combine this new interpretation with that of the λ, λ''s or a sphere, but I do not enter upon this, since it would lead us too far.

Clearly, the two angles of the system of co-ordinates:

$$\mathsf{A}_0=0,\ \mathsf{A}_1=0, \text{ and } \mathsf{A}_0=0,\ \mathsf{A}_2=0,$$

which appertain to $A=0$, are corresponding vanishing-points for f, for both remain unaltered for the collineation S [see *supra*, formula (41)]. Therefore $\mathsf{A}_0=0$ is one of the six straight lines which belong to f; $\mathsf{A}_1=0$, $\mathsf{A}_2=0$, the corresponding pole. In agreement herewith A_0 assumes, for our sixty substitutions, only the following twelve values, corresponding in pairs except as regards sign:

(11) $$\pm\mathsf{A}_0,\ \pm(\mathsf{A}_0+\epsilon^{\nu}\mathsf{A}_1+\epsilon^{4\nu}\mathsf{A}_2);$$

and in accordance with the same formula only the following five points are grouped with the point $\mathsf{A}_1=0$, $\mathsf{A}_2=0$:

(12) $$\mathsf{A}_0:\mathsf{A}_1:\mathsf{A}_2=1:2\epsilon^{4\nu}:2\epsilon^{\nu}.$$

I will describe the six points thus distinguished as *fundamental points* of the plane. If we connect a first fundamental point with the five others, we obtain the five straight lines:

$$\epsilon^{\nu}\mathsf{A}_1-\epsilon^{4\nu}\mathsf{A}_2=0.$$

Evidently the left sides of this equation are all contained as factors in the value of D just communicated. The curve $D=0$ must, however, be uniformly related to all the fundamental points. *Hence the curve $D=0$ splits up into the fifteen connecting lines of the six fundamental points.* The following algebraical decomposition corresponds to it:

(13) $$D=\prod_{\nu}(\epsilon^{\nu}\mathsf{A}_1-\epsilon^{4\nu}\mathsf{A}_2)\cdot\prod_{\nu}((1+\sqrt{5})\,\mathsf{A}_0+\epsilon^{\nu}\mathsf{A}_1+\epsilon^{4\nu}\mathsf{A}_2)$$
$$\cdot\prod_{\nu}((1-\sqrt{5})\,\mathsf{A}_0+\epsilon^{4\nu}\mathsf{A}_1+\epsilon^{\nu}\mathsf{A}_2),$$
$$(\nu=0,\ 1,\ 2,\ 3,\ 4),$$

as we easily verify. We could establish a number of interesting theorems about the fifteen straight lines here presenting themselves; they are the fifteen lines which belong to the point-pairs of T; they pass in threes through the ten points which we co-ordinated * to the point-pairs of H, &c. I do not

* Clebsch has incidentally dealt with the figure described in the text in the course of considerations allied, though again formulated quite differently, and has thus announced the last-mentioned property: *the six fundamental*

enter further here on these theorems, because we do not make further use of them; moreover, they are easily recognised as transferences of properties of the grouping of points which occur in connection with the icosahedron.

As regards the curves $B'=0$, $C'=0$, these have no special relation to our six fundamental points. *It is this circumstance which we will now make use of in order to replace B' and C' by two other expressions.* We shall introduce instead of B' a linear combination B of B' and A^3 in such wise that the curve $B=0$ contains the fundamental point $\mathsf{A}_1=0$, $\mathsf{A}_2=0$, and therefore (as an invariant curve) all the fundamental points. In the same way we shall replace C' by a linear combination C of C', A^2B, and A^5, which, equated to zero, represents a curve which has at $\mathsf{A}_1=0$, $\mathsf{A}_2=0$, and therefore at all the fundamental points a singular point of the highest possible kind. In this manner we find (after casting out particular numerical factors):

$$(14)\begin{cases} B=\dfrac{-B'+16A^3}{21}=8\mathsf{A}_0^4\mathsf{A}_1\mathsf{A}_2-2\mathsf{A}_0^2\mathsf{A}_1^2\mathsf{A}_2^2+\mathsf{A}_1^3\mathsf{A}_2^3-\mathsf{A}_0(\mathsf{A}_1^5+\mathsf{A}_2^5), \\ C=\dfrac{-C'-512A^5+1760A^2B}{187} \\ \quad =320\mathsf{A}_0^6\mathsf{A}_1^2\mathsf{A}_2^2-160\mathsf{A}_0^4\mathsf{A}_1^3\mathsf{A}_2^3+20\mathsf{A}_0^2\mathsf{A}_1^4\mathsf{A}_2^4+6\mathsf{A}_1^5\mathsf{A}_2^5 \\ \quad -4\mathsf{A}_0(\mathsf{A}_1^5+\mathsf{A}_2^5)(32\mathsf{A}_0^4-20\mathsf{A}_0^2\mathsf{A}_1\mathsf{A}_2+5\mathsf{A}_1^2\mathsf{A}_2^2)+\mathsf{A}_1^{10}+\mathsf{A}_2^{10}. \end{cases}$$

Evidently $B=0$ has at $\mathsf{A}_1=0$, $\mathsf{A}_2=0$, and thus at all the fundamental points not a merely ordinary point, but a double point, and is therefore (since we can show that it can possess no further double point) of deficiency 4. Similarly $C=0$ has at each of the fundamental points two cusps, *i.e.*, a 4-tuple point, and is therefore of deficiency $p=0$.

If we substitute in our new B, C, in accordance with formula (7):

$$\mathsf{A}_0=-\lambda_1\lambda_2,\quad \mathsf{A}_1=\lambda_2^2,\quad \mathsf{A}_2=-\lambda_1^2,$$

we have:

$$(15)\qquad B=-f(\lambda_1,\lambda_2),\quad C=-H(\lambda_1,\lambda_2),$$

which we may compare with (10). The relation which expresses

points form a tenfold Brianchon hexagon (Math. Ann., Bd. iv: "Ueber die Anwendung der quadratischen Substitution auf die Gleichungen 5. Grades und die geometrische Theorie des ebenen Fünfseits," 1871).

D^2 as an integral function of A, B, C, will therefore have the following terms not involving A:

$$D^2 = -1728B^5 + C^3.$$

Returning to the explicit values (9), (14), and taking account of a sufficient number of terms, we find the complete formula:*

$$(16) \quad D^2 = -1728B^5 + C^3 + 720ACB^3 - 80A^2C^2B + 64A^3(5B^2 - AC)^2.$$

§ 4. The Problem of the A's and its Reduction.

The Problem of the A's, as we proposed it, is fully determined by the explicit formulæ (6), (9), (14), now obtained for A, B, C, D, and the relation (16). We suppose the numerical values of A, B, C, D given in some manner in agreement with (16); *our problem requires us to determine the corresponding systems of values of* A_0, A_1, A_2. Since A, B, C, D form the full system of the invariant forms, our problem can only possess such solutions as proceed from some one thereof by the sixty substitutions (5). In fact, if we determine the number of solutions by Bezout's theorem, we shall be led to the number 60. Namely, from the values of A, B, C arise at the outset $2 \cdot 6 \cdot 10 = 120$ systems of values of the A's, of which, however, since A, B, C are all even functions of the A's, certain pairs can only differ by a simultaneous change of sign of the A's. Of these 120 systems of values, only half can therefore satisfy the given value of D, since D is of uneven order. All sixty systems of solution, as has already been said, proceed from some one thereof by the substitutions (5). *We can therefore say, in the sense explained previously* (I, 4), *that our problem is its own Galois resolvent after the adjunction of* ϵ, *and therefore possesses a group which is simply isomorphic with the group of the sixty icosahedral rotations.*

We consider now, in reliance on I, 5, § 4, the parallel *system of equations*. The ratios of $\mathsf{A}_0 : \mathsf{A}_1 : \mathsf{A}_2$ are evidently determined in sixty ways if in the equations:

$$(17) \qquad \frac{B}{A^3} = Y, \quad \frac{C}{A^5} = Z$$

* *Cf.* Brioschi, in tom. i of the Annali di Matematica (ser. 2, 1867), p. 228.

we can regard the values of Y and Z as known; * the required points **A** are the complete intersection of the curves of the sixth and tenth orders respectively:

$$B - Y \cdot A^3 = 0, \quad C - Z \cdot A^5 = 0.$$

From the sixty solutions of the equation system we now compute those of the corresponding form-problem rationally. Put, namely:

$$\frac{D}{A^7} = X. \tag{18}$$

Then if $\mathsf{A}_0 : \mathsf{A}_1 : \mathsf{A}_2 = a_0 : a_1 : a_2$ is one of the systems of solution of the equation problem, we have evidently:

$$\mathsf{A}_0 = \rho a_0, \quad \mathsf{A}_1 = \rho a_1, \quad \mathsf{A}_2 = \rho a_2, \tag{19}$$

understanding by ρ the following expression:

$$\rho = \frac{A^7(a_0, a_1, a_2)}{D(a_0, a_1, a_2)} \cdot X,$$

whereupon the statement is proved.

In this respect an essential difference exists between the binary form-problems previously studied and the present ternary ones; for then we required, as we showed in I, 3, § 2, one additional square root in the supplementary solution of the form-problem. This, of course, corresponds to the circumstance that the group of the homogeneous binary substitutions was only hemihedrically isomorphic with the non-homogeneous ones, while now simple isomorphism occurs. On the other hand, the two agree in another point. We could in the former case *reduce* the form-problem, as we called it, *i.e.*, replace the three magnitudes F_1, F_2, F_3 (connected by an equation of condition) on which the form-problem depended, by two independent variables, X and Y, which were themselves rational functions of F_1, F_2, F_3; whilst conversely the latter again depended rationally on them. We obtain just the same result in the problem of the **A**'s if we take into consideration the quotients X, Y, Z, which we just introduced in (17), (18). These magnitudes X, Y, Z, are in themselves defined as rational functions

* These magnitudes, Y, Z, are the same that we have denoted by a, b, in II, 1, § 7 [formula (36)].

of A, B, C, D, but we can conversely also express A, B, C, D rationally by means of X, Y, Z. In fact, if we divide in (16) both terms by A^{14}, we have, after an easy rearrangement, in virtue of (17), (18):

$$(20)\qquad A=\frac{X^2}{Z^3-1728Y^5+720Y^3Z-80YZ^2+64(5Y^2-Z)^2},$$

while

$$(21)\qquad B=Y\,.\,A^3,\quad C=Z\,.\,A^5,\quad D=X\,.\,A^7,$$

which are the desired formulæ.

It is instructive to bring forward also for comparison here the problem of the y_ν's, which we studied in the preceding chapter as the canonical equation of the fifth degree. We then supposed that, in addition to the coefficients α, β, γ, of the equation, the square root ∇ of the discriminant was given, the square of which is an integral function of α, β, γ. We then obtained sixty systems of solution y_0, y_1, y_2, y_3, y_4, which were again fully determined (rationally) in terms of the corresponding values of the ratios $y_0:y_1:y_2:y_3:y_4$. This depends on the fact that we can construct as before from the given magnitudes quotients $\left(e.g., \frac{\beta}{\alpha} \text{ or } \frac{\gamma}{\beta}\right)$ which are of the first dimension in the y's. We can also reduce the form-problem of the y's, only this is not so simply attained as in the other cases. *The reduction is actually given by the m, n, Z, of the canonical resolvent of the icosahedron.* We have represented, viz., in I, 4, § 12, § 14, α, β, γ, ∇, rationally in terms of m, n, Z, while conversely we have just now (in II, 3) given exhaustive methods in virtue of which m, n, Z appear as rational functions of α, β, γ, ∇.

If $A=0$ for the form-problem of A's, we can solve it directly by means of the icosahedral equation:

$$\frac{H^3(\lambda_1,\lambda_2)}{1728f^5(\lambda_1,\lambda_2)}=\frac{C^3}{1728B^5},$$

namely, if we have determined $\lambda_1:\lambda_2$ from it, we find by formula (7):

$$\mathsf{A}_0:\mathsf{A}_1:\mathsf{A}_2=-\lambda_1\lambda_2:\lambda_2^2:-\lambda_1^2,$$

and hence, as we saw above [formula (19)], the values of A_0, A_1, A_2 themselves.

§ 5. On the Simplest Resolvents of the Problem of the **A**'s.

We will now consider the simplest resolvents of the problem of the **A**'s. It is evident, from what we know of the group of the problem, that we shall have to deal here with resolvents of the fifth and sixth degree. Our problem will be merely to establish the *simplest* rational and integral functions respectively of the **A**'s, which assume, for the substitutions known to us, five and six values respectively. *The principle of transference developed in* § 2 *here again serves our purpose: we take the simplest integral functions of* λ_1, λ_2, *which assume for the homogeneous icosahedral substitutions five or six values, polarise these with respect to* λ_1', λ_2', *till a function arises which is symmetrical in the* λ, λ''*s, and finally substitute the* **A**'*s in place of the symmetrical combinations of the* λ, λ''*s.*

As regards the five-valued functions of the λ_1, λ_2's, the simplest were:

$$(22)\quad \begin{aligned} t_\nu(\lambda_1, \lambda_2) &= \epsilon^{3\nu}\lambda_1^6 + 2\epsilon^{2\nu}\lambda_1^5\lambda_2 - 5\epsilon^\nu\lambda_1^4\lambda_2^2 \\ &\quad - 5\epsilon^{4\nu}\lambda_1^2\lambda_2^4 - 2\epsilon^{3\nu}\lambda_1\lambda_2^5 + \epsilon^{2\nu}\lambda_2^6, \\ W_\nu(\lambda_1, \lambda_2) &= -\epsilon^{4\nu}\lambda_1^8 + \epsilon^{3\nu}\lambda_1^7\lambda_2 - 7\epsilon^{2\nu}\lambda_1^6\lambda_2^2 - 7\epsilon^\nu\lambda_1^5\lambda_2^3 \\ &\quad + 7\epsilon^{4\nu}\lambda_1^3\lambda_2^5 - 7\epsilon^{3\nu}\lambda_1^2\lambda_2^6 - \epsilon^{2\nu}\lambda_1\lambda_2^7 - \epsilon^\nu\lambda_2^8; \end{aligned}$$

to them are added, further, t_ν^2 and $t_\nu W_\nu$. Now, polarising t_ν thrice, W_ν four times, and introducing the **A**'s, we obtain accordingly as the *simplest five-valued function of the* **A**'*s:*

$$(23)\quad \left\{ \begin{aligned} \delta_\nu = {} & \epsilon^\nu(4A_0^2A_2 - A_1A_2^2) + \epsilon^{2\nu}(-2A_0A_2^2 + A_1^3) \\ & + \epsilon^{3\nu}(2A_0A_1^2 - A_2^3) + \epsilon^{4\nu}(-4A_0^2A_1 + A_1^2A_2), \end{aligned} \right.$$

$$\left\{ \begin{aligned} \delta_\nu' = {} & \epsilon^\nu(-4A_0^3A_2 + 3A_0A_1A_2^2 - A_1^4) \\ & + \epsilon^{2\nu}(-6A_0^2A_2^2 + A_0A_1^3 + A_1A_2^3) \\ & + \epsilon^{3\nu}(-6A_0^2A_1^2 + A_0A_2^3 + A_1^3A_2) \\ & + \epsilon^{4\nu}(-4A_0^3A_1 + 3A_0A_1^2A_2 - A_2^4); \end{aligned} \right.$$

if we want more five-valued functions, we shall take in addition δ_ν^2 and $\delta_\nu\delta_\nu'$, corresponding to t_ν^2 and $t_\nu W_\nu$. The resolvent of the δ 's we shall presently discuss more in detail.

Of the resolvents of the sixth degree of the icosahedral equation, we have previously (I, 5, § 15) only considered the one whose roots ϕ are given by the formulæ:

$$(24) \qquad \begin{cases} \phi_\infty = 5\lambda_1{}^2\lambda_2{}^2, \\ \phi_\nu = (\epsilon^\nu\lambda_1{}^2 + 2\lambda_1\lambda_2 - \epsilon^{4\nu}\lambda_2{}^2)^2. \end{cases}$$

We obtain from this by our principle of transference the following roots of a resolvent of the sixth degree of the A's:

$$(25) \qquad \begin{cases} z_\infty = 5\mathsf{A}_0{}^2, \\ z_\nu = (\epsilon^\nu\mathsf{A}_2 + \mathsf{A}_0\epsilon^{4\nu} \quad {}_1)^2. \end{cases}$$

Here, however, we have exactly the equations of definition of the Jacobian equations of the sixth degree given in II, 1, § 3; at most we should have to mark the distinction that here ϵ^ν stands where $\epsilon^{4\nu}$ stood before and conversely. But this is merely a difference in the denomination of the roots z_ν. If we recur to the formulæ which we also communicated, *loc. cit.*, § 5, in describing the Jacobian equations, we learn first that our present magnitudes A, B, C are exactly identical with those similarly denoted there. We can therefore, without more ado, carry over the form of the Jacobian equation previously communicated:

$$(26) \quad (z-A)^6 - 4A(z-A)^5 + 10B(z-A)^3 - C(z-A) + (5B^2 - AC) = 0;$$

the only question is on what basis we shall place it from our present standpoint. The further question arises how far we can replace the problem of the A's by the equation (26), and, in particular, what significance is then to be attributed to our form D.

§ 6. The General Jacobian Equation of the Sixth Degree.

We have already in formula (11) come across the linear functions of the A's whose squares represent the roots z (25) of the Jacobian equation of the sixth degree; we there saw that these, when equated to zero, represent the *polars of the six fundamental points with respect to the conic* $A=0$, and therefore certain straight lines which do not themselves perhaps pass through the fundamental points. We can, however, introduce in their place curves which do so; namely, we recognise at once that the conics:

$$z_\nu - A = 0 \ (\nu = 0, 1, 2, 3, 4)$$

all pass through $\mathsf{A}_1 = 0$, $\mathsf{A}_2 = 0$, *and that therefore of the conics:*

$$z_\infty - A = 0, \quad z_\nu - A = 0,$$

each contains those five fundamental points whose indices are different from its own. We will now consider the $(z-A)$'s as the actual unknowns. Then the theorem just given permits us to write down at once (paying regard to the definition of B, C contained in § 3) the coefficients of the corresponding equation, so far as their form is concerned. Let us consider, for example, the sum:

$$\Sigma(z_i - A)\,(z_k - A)\,(z_l - A)$$

(where the summation extends over all the values of i, k, l which differ from one another), which will give the third coefficient of that equation: it must be equal to an invariant form of the **A**'s of the sixth degree which vanishes twice for all the fundamental points, and can there only differ from B by a numerical factor. In this way we obtain directly:

$$(z-A)^6 + kA(z-A)^5 + lB(z-A)^3 + mC(z-A) + (nB^2 + pAC) = 0,$$

where k, l, m, n, p are numerical coefficients which are as yet unknown, but which we determine readily hereafter by returning to the explicit values of the expressions in the **A**'s which present themselves. The coincidence with formula (26) is obvious. Let us further remark that (26) is in fact transformed into the resolvent of the sixth degree (which we previously established) of the icosahedral equation, if we put in agreement with (24) and (15):

$$A = 0, \quad B = -f, \quad C = -H, \quad z = \phi.$$

As regards the *group* of the equation (26) [in the Galois sense], this is determined already by our earlier elucidation of the case $A=0$, to which we here refer (I, 4, § 15). It is a group of sixty permutations which is simply isomorphic with the group of substitutions of the **A**'s. It must therefore be possible to express the **A**'s rationally by means of our z's. We effect this most simply if we first compute from the equations (25) the squares of the **A**'s and the products in twos, and hence derive the quotients $\mathbf{A}_0 : \mathbf{A}_1 : \mathbf{A}_2$, and then proceed exactly as in § 4. Here we must manifestly make use of the D in addition to A, B, C, which alone appear in the coefficients of (26). We can therefore say:

The Jacobian equation (26) *is an equivalent of the problem of the* **A**'*s, if we suppose that, besides its coefficients, D is also given,* i.e. (*by* (16)): *the square-root of a definite integral function of A, B, C.*

We now ask how D^2 may be expressed as a rational function of the roots z. To this end we form from (25) the difference of any two z's as a function of the **A**'s, and find that this, being a difference of two squares, can, after separation of a constant factor, be also split up into linear factors, such as, by formula (13), also appear in D. We get, for example:

$$z_\nu - z_{2\nu} = (\epsilon^\nu - \epsilon^{4\nu})\,(\epsilon^\nu \mathsf{A}_1 - \epsilon^{4\nu}\mathsf{A}_2)\,\big((1 \pm \sqrt{5})\,\mathsf{A}_0 + \epsilon^\nu \mathsf{A}_1 + \epsilon^{4\nu}\mathsf{A}_2\big)$$

for $\nu = 1, 2, 3, 4$, where $+\sqrt{5}$ is to be taken for $\nu = 2, 3$; $-\sqrt{5}$ for $\nu = 1, 4$. If we now multiply all these differences together (each taken once), we obtain on the left-hand side the square root of the discriminant of (26), which we have already (II, 1, § 5) denoted by Π. On the right hand, however, the constant factors give $\pm\sqrt{5^5}$, the rest just D^2, so that therefore:

$$D^2 = \sqrt{\frac{\Pi}{5^5}}, \quad \text{or} \quad D = \sqrt[4]{\frac{\Pi}{5^5}}. \tag{27}$$

Here D appears, as we see, in the form of an *accessory* irrationality, *i.e.*, as an irrational function of the z's. This will not be the case if, with Herr Kronecker, we regard not the z's, but the $\sqrt{z}$'s as the unknowns of (26); for we can immediately express A_0, A_1, A_2 linearly in terms of the $\sqrt{z}$'s. But even then the statement of the problem is not fixed by (26) alone, but the value of D must be given expressly besides. I believe, therefore, that it is not to the purpose to make the Jacobian equations of the sixth degree the keystone of the theory, but that it is better to begin, as we have done, with the problem of the **A**'s as such.

§ 7. Brioschi's Resolvent.

We follow yet further the connection of our considerations with the developments of Brioschi and Kronecker by now studying, first of all, that simplest resolvent of the fifth degree whose roots are the expressions δ_ν (23). *This must give exactly*

Brioschi's resolvent, of which we gave an account in II, 1, § 5. For the δ_ν's are completely identical, as an actual comparison teaches us, with what was then denoted [formula (22)] by x_ν.

In order to compute our equation of the fifth degree, we first seek, as before, the geometrical significance of the δ_ν's. We remark at the outset that all the δ_ν's vanish for $\mathsf{A}_1=0$, $\mathsf{A}_2=0$. They therefore represent, when equated to zero, curves of the third order which pass through all the fundamental points. But more than this: the product of the δ_ν's must be, as an invariant form of the fifteenth degree in the A's, identical with D save as to a factor, while $D=0$ represents, as we know, the fifteen connecting lines of the six fundamental points. *Hence each of the δ_ν's, when equated to zero, represents three straight lines, which taken together contain the whole set of fundamental points.* We verify accordingly the following decomposition:

$$(28) \qquad \delta_\nu=(\epsilon^{4\nu}\mathsf{A}_1-\epsilon^\nu\mathsf{A}_2)\,.\,\big((1+\sqrt{5})\mathsf{A}_0+\epsilon^{4\nu}\mathsf{A}_1+\epsilon^\nu\mathsf{A}_2\big)\cdot\big((1-\sqrt{5})\mathsf{A}_0+\epsilon^{4\nu}\mathsf{A}_1+\epsilon^\nu\mathsf{A}_2\big).$$

We conclude therefrom that the product $\delta_0\delta_1\delta_2\delta_3\delta_4$ is actually identical with D (not merely to a factor *près*). As regards the other symmetric functions of the δ's, we have in any case:

$$\Sigma\delta=0, \quad \Sigma\delta^3=0,$$

for there are no invariant forms of the third or ninth degree. We further conclude from the relation of δ to the fundamental points that:

$$\Sigma\delta^2=kB, \quad \Sigma\delta^4=lB^2+mAC,$$

understanding by k, l, m, appropriate numerical factors. On determining the latter we have finally:

$$(29) \qquad \delta^5+10B\,.\,\delta^3+5(9B^2-AC)\delta-D=0,$$

agreeing with Brioschi,* and agreeing further with the special formula which we derived in I, 4, § 11, on the supposition that $A=0$. The discriminant of (29) is of course a complete square.

* In Brioschi's memoir somewhat different numerical coefficients were originally given, but these were afterwards rectified by Herr Joubert: "Sur l'équation du sixième degré," Comptes Rendus, t. 64 (1867, 1); see in particular pp. 1237–1240.

There is no difficulty in computing the product $\prod_{\nu<\nu'}(\delta_\nu - \delta_{\nu'})$ as an integral function of A, B, C. For $A=0$ it will become $-25\sqrt{5}\,.\,C^3$, by I, 4, § 14.

The equation (29) is necessarily the more interesting, because it represents, in the sense of our previous terminology, *the general diagonal equation of the fifth degree.* To express it geometrically, we can say that the formulæ (23) for δ_ν, inasmuch as they satisfy identically the relation $\Sigma\delta=0$, $\Sigma\delta^3=0$, *give a single-valued representation of the diagonal surface on the plane* **A**. This representation is a special case of that well-known one which was given * by *Clebsch* and *Cremona* for general surfaces of the third order, and which *Clebsch* has studied for the diagonal surface just in the form here in question.† For to the plane sections of the diagonal surface correspond in general, in virtue of (23), such curves of the third order as intersect one another in the six fundamental points of the plane which now *become the fundamental points of the representation.* Here the intersection of the diagonal surface with the canonical surface is represented by $B=0$ (as follows from (29)), while the curves $A=0$, $C=0$, taken together, represent those two twisted curves of the sixth order on the diagonal surface which are the geometrical locus of points with the pentahedral co-ordinates t_ν (II, 3, § 4). This is in accordance with the fact that we have, in § 3 of the present chapter, found the deficiency p of the curves $B=0$, $A=0$, $C=0$, equal to 4, 0, 0.

§ 8. Preliminary Remarks on the Rational Transformation of our Problem.

Of the researches mentioned above relating to Jacobian equations of the sixth degree, those still remain which relate to the problem: from a first Jacobian equation of the sixth degree to establish a second by a transformation rational in the $\sqrt{z}$'s

* *Cf.* Salmon-Fiedler, "Analytische Geometrie des Raumes," 3d edition, 1879–80.

† Viz., in the memoir just cited: "Ueber die Anwendung der quadratischen Substitution auf die Gleichungen 5. Grades," Math. Annalen, Bd. iv (1871).

and as general as possible. I will expound these researches from our own standpoint, without entering further into the historical relations thereof. Our object is *to determine three magnitudes,* B_0, B_1, B_2, *in as general a manner as possible, as rational homogeneous functions of the* A_0, A_1, A_2, *in such wise that they themselves undergo the linear substitutions of* § 2 *when we subject* A_0, A_1, A_2, *to the same.**

Our requirement, be it understood, by no means demands that the individual substitution of the B's should be *identical* with that of the A's; it is only necessary that the totality of the substitution should be mutually coincident. We know, so far, two possibilities of attaining such coincidence: first, by making the substitutions of the B's actually identical with those of the A's; secondly, by allowing them to proceed from the substitutions of the A's on writing ϵ^2 in place of ϵ throughout; in the first case we speak of *cogredient,* in the second case of *contragredient* variables. In the next paragraph but one I shall show how we can thus arrive at a separation of the two cases *a priori,* and that, besides them, no others possessing individual importance can exist. Meanwhile let us take our cases as given empirically, and ask how they are substantiated by definite formulæ.

It will be to the purpose to first deal with the corresponding statement of the problem in the domain of binary variables, where we came repeatedly into contact with them in our earlier chapters. Let κ_1, κ_2 be homogeneous rational, not necessarily integral, functions of λ_1, λ_2:

$$(30) \qquad \kappa_1 = \phi_1(\lambda_1, \lambda_2), \quad \kappa_2 = \phi_2(\lambda_1, \lambda_2),$$

we require so to determine ϕ_1, ϕ_2 that κ_1, κ_2 are either cogrediently or contragrediently transformed when λ_1, λ_2 are subjected to the homogeneous icosahedral substitutions. To this end we construct the form, binary in two sets of variables:

$$(31) \qquad F(\lambda_1, \lambda_2; \mu_1, \mu_2) = \mu_1 \cdot \phi_2(\lambda_1, \lambda_2) - \mu_2 \cdot \phi_1(\lambda_1, \lambda_2).$$

This evidently remains invariant if we subject λ_1, λ_2 to the original icosahedral substitutions, μ_1, μ_2, to the co-ordinated ones

* Our demand for entirely *homogeneous* functions is, one may say, an unnecessary restriction, and one which we can afterwards remove, but which we will retain in our exposition in order to be able to employ our geometrical phraseology. See the analogous remark in II, 2, § 7.

(cogredient or contragredient); for it is equal to $\mu_1\kappa_2-\mu_2\kappa_1$; and μ_1, μ_2 and κ_1, κ_2 undergo in each case identical substitutions of determinant unity. Conversely, if we have a form in the λ, μ's invariant in this sense, and which is linear in the μ's and homogeneous in λ_1, λ_2, then:

$$\kappa_1 = -\frac{\delta F}{\delta \mu_2}, \quad \kappa_2 = \frac{\delta F}{\delta \mu_1} \tag{32}$$

will be a solution of the problem proposed. *It simply comes to this, therefore: to establish all invariant forms F.*

Now let us observe the following facts. If we have found two systems of solution of (30) κ_1, κ_2; κ_1', κ_2', the determinant $\kappa_1\kappa_2'-\kappa_2\kappa_1'$ remains invariant for all the icosahedral substitutions. But this is equal to the functional determinant of the corresponding forms F, F':

$$\begin{vmatrix} \frac{\delta F}{\delta \mu_1} & \frac{\delta F}{\delta \mu_2} \\ \frac{F'}{\delta \mu_1} & \frac{\delta F'}{\delta \mu_2} \end{vmatrix},$$

and this, therefore, as a rational function of λ_1, λ_2, must be a rational function of the icosahedral forms $f(\lambda_1, \lambda_2)$, $H(\lambda_1, \lambda_2)$, $T(\lambda_1, \lambda_2)$. I will now assume that we know some two of the required forms F_1, F_2, with a non-evanescent functional determinant. Then, if we apply the identity:

$$\begin{vmatrix} F & F_1 & F_2 \\ \frac{\delta F}{\delta \mu_1} & \frac{\delta F_1}{\delta \mu_1} & \frac{\delta F_2}{\delta \mu_1} \\ \frac{\delta F}{\delta \mu_2} & \frac{\delta F_1}{\delta \mu_2} & \frac{\delta F_2}{\delta \mu_2} \end{vmatrix} = 0,$$

it follows, from the theorem just established, that each of the forms we seek is compounded of F_1, F_2, in the following form:

$$F = R_1 \cdot F_1 + R_2 \cdot F_2, \tag{33}$$

where R_1, R_2 are rational functions of $f(\lambda_1, \lambda_2)$, $H(\lambda_1, \lambda_2)$, $T(\lambda_1, \lambda_2)$. But, conversely, if we assume R_1, R_2 to be rational functions of this kind, and then only take account of the rule that F is to be homogeneous in λ_1, λ_2, F will be a form of the kind required. *Hence* (33) *contains in general the solution of our problem, provided only we regard two of our forms* F_1, F_2 *as known.* But this supposition is, in fact, admissible both in the contragredient and the cogredient case. Indeed, we know in both cases the lowest forms F_1, F_2, *i.e.*, those whose degree in λ_1, λ_2 is as low as possible. In the contragredient case these are the two forms N_1, M_1, which we always employed in the preceding chapter:

$$(34)\quad \begin{cases} F_1 = N_1 = \mu_1(7\lambda_1^5\lambda_2^2 + \lambda_2^7) + \mu_2(-\lambda_1^7 + 7\lambda_1^2\lambda_2^5), \\ F_2 = M_1 = \mu_1(\lambda_1^{13} - 39\lambda_1^8\lambda_2^5 - 26\lambda_1^3\lambda_2^{10}) \\ \qquad + \mu_2(26\lambda_1^{10}\lambda_2^3 - 39\lambda_1^5\lambda_2^8 - \lambda_2^{13}), \end{cases}$$

while in the cogredient case we have the two following:

$$(35)\quad \begin{cases} F_1 = \lambda_2\mu_1 - \lambda_1\mu_2, \\ F_2 = \dfrac{\delta f}{\delta\lambda_1}\cdot\mu_1 + \dfrac{\delta f}{\delta\lambda_2}\cdot\mu_2. \end{cases}$$

Thus the question we raised is completely solved, so far as the domain of binary forms is concerned.*

§ 9. Accomplishment of the Rational Transformation.

Returning now to the **A**'s, we can begin in their case with a step which is analogous to the transition from (30) to (31); in other words, instead of seeking elements B_0, B_1, B_2, which are *covariant* to A_0, A_1, A_2, in the one sense or the other, we seek an *invariant* which contains simultaneously both sets of variables. The feasibility of this is, geometrically speaking, founded on the fact that an invariable conic:

$$\mathsf{B}_0^2 + \mathsf{B}_1\mathsf{B}_2 = 0$$

lies in the plane **B**, and that, in respect to this conic, to every point B_0, B_1, B_2, there is co-ordinated as a covariant a certain straight line, to wit, the corresponding polar:

$$2\mathsf{B}_0 \,.\, \mathsf{A}_0' + \mathsf{B}_2 \,.\, \mathsf{A}_1' + \mathsf{B}_1 \,.\, \mathsf{A}_2' = 0.\dagger$$

If, therefore, the following formulæ:

$$(36)\quad \mathsf{B}_0 = \phi_0(\mathsf{A}_1, \mathsf{A}_2, \mathsf{A}_3),\ \mathsf{B}_1 = \phi_1(\mathsf{A}_0, \mathsf{A}_1, \mathsf{A}_2),\ \mathsf{B}_2 = \phi_2(\mathsf{A}_0, \mathsf{A}_1, \mathsf{A}_2)$$

* As regards the contragredient case, we have already become acquainted, in formula (25) of II, 3, § 4, with a particular case which is included in this solution.

† Here A_0', A_1', A_2' denote the current point co-ordinates.

co-ordinate the **B**'*s to the* **A**'*s either as cogredients or as contragredients, the form derived from them:*

(37) $F(\mathsf{A}_0, \mathsf{A}_1, \mathsf{A}_2 ; \mathsf{A}_0', \mathsf{A}_1', \mathsf{A}_2') = 2\phi_0 \cdot \mathsf{A}_0' + \phi_2 \cdot \mathsf{A}_1' + \phi_1 \cdot \mathsf{A}_2'$,

will be invariant, provided we effect the same substitutions on the **A**″*s as on the* **B**'*s. Conversely, provided F is an invariant in the sense explained:*

(38) $$\mathsf{B}_0 = \frac{1}{2} \cdot \frac{\delta F}{\delta \mathsf{A}_0'}, \quad \mathsf{B}_1 = \frac{\delta F}{\delta \mathsf{A}_2'}, \quad \mathsf{B}_2 = \frac{\delta F}{\delta \mathsf{A}_1'}$$

are formulæ of the nature which we are seeking.

We now remark that every F admits of being composed with three such F's, which are linearly independent, in the form:

(39) $$F = R_1 F_1 + R_2 F_2 + R_3 F_3,$$

where R_1, R_2, R_3 are rational functions of the invariant forms, which depend only on A_0, A_1, A_2, *i.e.*, rational functions of A, B, C, D. Conversely, if we take R_1, R_2, R_3 as rational functions of this kind, we shall always obtain from (39) a form F of the nature we desire, where we have it in our power, if we attach importance thereto, to make F a homogeneous function of A_0, A_1, A_2. *Everything is therefore reduced to finding, in two cases, three forms,* F_1, F_2, F_3, *of as low degree in the* **A**'*s as possible.*

In the case of cogredients we solve this problem directly by the construction of polars, a process to which we subject the lowest invariant forms, which only contain **A**'s, *i.e.*, A, B, C. *We shall put, namely:*

(40) $$\begin{cases} F_1 = 2\mathsf{A}_0 \cdot \mathsf{A}_0' + \mathsf{A}_2 \cdot \mathsf{A}_1' + \mathsf{A}_1 \cdot \mathsf{A}_2', \\ F_2 = \dfrac{\delta B}{\delta \mathsf{A}_0} \cdot \mathsf{A}_0' + \dfrac{\delta B}{\delta \mathsf{A}_1} \cdot \mathsf{A}_1' + \dfrac{\delta B}{\delta \mathsf{A}_2} \cdot \mathsf{A}_2', \\ F_3 = \dfrac{\delta C}{\delta \mathsf{A}_0} \cdot \mathsf{A}_0' + \dfrac{\delta C}{\delta \mathsf{A}_1} \cdot \mathsf{A}_1' + \dfrac{\delta C}{\delta \mathsf{A}_2} \cdot \mathsf{A}_2'. \end{cases}$$

In the case of contragredients, on the other hand, we again recur to the principle of transference of § 2. We shall first obtain three forms:

$$\Omega(\lambda_1, \lambda_2 ; \mu_1, \mu_2)$$

invariant for contragredient icosahedral substitutions, which are of even degree $2n$, of the second degree in the μ's, and of the lowest possible degree in the λ's. *Then we shall polarise these* Ω*'s n-times with respect to* λ, *introducing* λ' *by the operation, and once with respect to* μ, *introducing* μ', *and shall finally replace the symmetric functions of the* λ, λ'*'s by the* A*'s, those of the* μ, μ'*'s by the* A$'$*'s, writing therefore:*

$$(41)\quad \begin{cases} A_0 = -\frac{1}{2}(\lambda_1\lambda_2' + \lambda_2\lambda_1'), \quad A_1 = \lambda_2\lambda_2', \quad A_2 = -\lambda_1\lambda_1'; \\ A_0' = -\frac{1}{2}(\mu_1\mu_2' + \mu_2\mu_1'), \quad A_1' = \mu_2\mu_2', \quad A_2' = -\mu_1\mu_1'.^* \end{cases}$$

The forms Ω which are here most suited to our purpose we can borrow from the data of Herr Gordan previously cited. As the Ω_1, we choose the form τ, which we have communicated in § 11 of the preceding chapter (formula (60)):

$$(42)\quad \Omega_1 = \mu_1^2(-\lambda_1^6 - 3\lambda_1\lambda_2^5) + 10\mu_1\mu_2 \cdot \lambda_1^3\lambda_2^3 + \mu_2^2(3\lambda_1^5\lambda_2 - \lambda_2^6).$$

In order, then, to obtain Ω_2, we construct the functional determinant of the ground-form a noted in that chapter, and the N_1 similarly employed just now:

$$\begin{vmatrix} \frac{\delta a}{\delta\mu_1} & \frac{\delta a}{\delta\mu_2} \\ \frac{\delta N_1}{\delta\mu_1} & \frac{\delta N_1}{\delta\mu_2} \end{vmatrix}.$$

We thus obtain:

$$(43)\quad \Omega_2 = \mu_1^2(-10\lambda_1^8\lambda_2^2 + 20\lambda_1^3\lambda_2^7) + 2\mu_1\mu_2(-\lambda_1^{10} + 14\lambda_1^5\lambda_2^5 + \lambda_2^{10}) + \mu_2^2(-20\lambda_1^7\lambda_2^3 - 10\lambda_1^2\lambda_2^8).$$

Finally, we bring forward as the Ω_3 the square of N_1:

$$(44)\quad \Omega_3 = [\mu_1(-7\lambda_1^5\lambda_2^2 - \lambda_2^7) + \mu_2(\lambda_1^7 - 7\lambda_1^2\lambda_2^5)]^2.$$

Now, applying our process of transformation first to Ω_1, there arises—disregarding a numerical factor—the following as the simplest form F_1, of the third degree in A_0, A_1, A_2:

$$(45)\quad F_1 = 2A_0'(2A_0^3 - 3A_0A_1A_2) - A_1'(3A_0A_2^2 + A_1^3) - A_2'(3A_0A_1^2 + A_2^3).$$

We now treat Ω_2 (43) in a similar manner, but subtract from

* Of course we could also proceed in just the same way in the case of cogredients; we should not, however, obtain any results different from those now communicated, and should only have to repeat once more the process of polarisation which led us above to A, B, C, D.

the result, for the sake of simplification, a certain multiple of $A \cdot F_1$. Thus we have:

(46) $$F_2 = 2A_0'(-8A_0^3A_1A_2 + 6A_0A_1^2A_2^2 - A_1^5 - A_2^5) + A_1'(16A_0^3A_2^2 - 8A_0^2A_1^3 - 4A_0A_1A_2^3 + 2A_1^4A_2) + A_2'(16A_0^3A_1^2 + 8A_0^2A_2^3 - 4A_0A_1^3A_2 + 2A_1A_2^4).$$

We finally deal with Ω_3 (44), and obtain, after subtracting proper multiples of $A^2 \cdot F_1$ and $A \cdot F_2$:

(47) $$F_3 = 2A_0'(32A_0^3A_1^2A_2^2 - 4A_0^2(A_1^5 + A_2^5) - 16A_0A_1^3A_2^3 + 3A_1A_2(A_1^5 + A_2^5)) + A_1'(-32A_0^5A_2^2 + 48A_0^4A_1^3 - 32A_0^3A_1A_2^3 - 4A_0^2A_1^4A_2 + 14A_0A_1^2A_2^4 - 3A_1^5A_2^2 - A_2^7) + A_2'(-32A_0^5A_1^2 + 48A_0^4A_2^3 - 32A_0^3A_1^3A_2 - 4A_0^2A_1A_2^4 + 14A_0A_1^4A_2^2 + 3A_1^2A_2^5 - A_1^7).$$

On introducing the F_1, F_2, F_3 thus obtained in (39), and through this in (38), our task is completely accomplished in the contragredient case also.

§ 10. Group-Theory Significance of Cogredience and Contragredience.

We now return to the group-theory question, to which we were led at the beginning of § 8. The linear substitutions of the **B**'s are, at all events, simply isomorphic with those of the **A**'s; we have finally to deal with the problem of investigating in how many different ways the group of sixty icosahedral substitutions:

(48) $$V_0,\ V_1,\ \ldots\ V_{59}$$

can be co-ordinated to itself in simple isomorphism. Two sorts of this co-ordination are given by cogredience and contragredience; we will show that all others are essentially reduced to these.

I must state at the outset what rearrangements of (48) will be regarded as non-essential. They are those rearrangements which arise from *transformation* in the sense previously explained (I, 1, § 2), which, therefore, replace any V_k by $(V')^{-1}V_kV'$, where by V' is to be understood any operation of (48). In the applications, namely, which we have to make, we can always regard such a rearrangement as a mere change of the system of co-ordinates. If we replace the variable z which is subjected to the icosahedral substitutions (48) by $z' = V'(z)$, $(V')^{-1}V_k V'$ will appear throughout in place of V_k; and similarly if we regard the V_k's as the ternary substitutions of $\mathbf{A}_0$, $\mathbf{A}_1$, $\mathbf{A}_2$.

With the intention of again applying the "principle of transformation" just formulated, we now recur to the generation of the icosahedral group from two operations S and T, of which the first has the period 5, the second the period 2 (I, 1, § 12). We shall have determined the co-ordination which we seek as soon as we declare what operations S', T', are to correspond to S, T. Here S' will in any case have to possess the period 5. But, by I, 1, § 8, there are in the icosahedral group 24 operations altogether of period 5, of which 12 are associated with S, the other 12 with S^2. If, therefore, in the co-ordination which we are seeking, we call to our aid a modification of these by an appropriate transformation of the group, *we can in every case put S' equal either to S or S^2.* If this is done, S' remains unaltered when we replace V_k in general by $S^{-\nu} V_k S^{\nu}$ ($\nu = 0, 1, 2, 3, 4$). Consider now the fifteen operations of period 2 which are contained in (48). If we choose ν properly in the transformation just mentioned, we can always reduce an individual operation of period 2 to one of the three following:

$$T, \quad TU, \quad U,$$

where U is defined as in I, 1, § 8 (compare I, 2, § 6). *If, therefore, we have disposed S' in the manner just mentioned, it is sufficient to make T' equal to one of the three operations T, TU, U.* Compare now the rules of periodicity in I, 2, § 6. In accordance with them, ST has the period 3, therefore $S'T'$ must also have the period 3. But now we find in the same place for ST, STU, SU, S^2T, S^2TU, S^2U respectively, the periods 3, 5, 2, 5, 3, 2 assigned. Hence $S'T'$ can only be either ST or S^2TU. *There remain, therefore, but two possibilities: in the one case we put $S' = S$, $T' = T$, in the other $S' = S^2$, $T' = TU$.* If we write down the corresponding icosahedral substitutions, we recognise that S^2 and TU emanate from S and T when we change ϵ into ϵ^2. *Thus we are, in fact, brought back to just the two cases, cogredience and contragredience*, as was to be proved.

We can evidently repeat for every group the question which is thus answered for the case of the icosahedron. If, then, a form-problem is proposed which belongs to a group already investigated, we can demand algebraical developments corresponding to those given in §§ 8, 9. I will not enter here on a general exposition of this, which would lead us beyond our subject (see, however, I, 5, § 5). I will only remark that the case of cogredience (which, of course, always exists) can always be solved by the construction of polars, when among the invariants of the form-problem there is one of the second degree. This occurs especially in those form-problems of which the variables $x_0, x_1 \ldots x_{n-1}$ are simply permuted, and which are therefore represented by equations of the nth degree with unconditioned coefficients. If for these we employ the invariant Σx^2 in just the same way as we applied the conic $\mathbf{B}_0^2+\mathbf{B}_1\mathbf{B}_2$ just now (§ 9), we are in a position to make the differential coefficients $\frac{\delta\phi}{\delta x_0}, \frac{\delta\phi}{\delta x_1}, \ldots \frac{\delta\phi}{\delta x_{n-1}}$ covariant to $x_0, x_1 \ldots x_{n-1}$, where by ϕ is to be understood any form which is invariant with respect to the permutations of the group. We are evidently led back, as a consequence of this method, to exactly the *transformation of Tschirnhaus* when we take into consideration, in particular, as the functions ϕ the sums of powers of the x's. The old process of Tschirnhaus is therefore, together with formulæ (38), embraced by a general method relating to form-problems of a certain class. Compare with this what was said in II, 2, § 7, on the co-ordination of points and planes.*

§ 11. Introductory to the Solution of our Problem.

We will maintain for a moment the analogy with the Tschirnhausian transformation, and accordingly consider the coefficients R_1, R_2, R_3 in (39) as undetermined magnitudes. If we then compute for the corresponding $\mathbf{B}_0, \mathbf{B}_1, \mathbf{B}_2$, the expression $\mathbf{B}_0^2+\mathbf{B}_1\mathbf{B}_2$, we obtain a quadratic form of these

* We can generalise somewhat the remark in the text. In order that the construction of polars may aid in the attainment of our object, it is not necessary that an invariant form quadratic in the x's should exist; the presence of an invariant form *bilinear in the* x, x''s is sufficient. In this sense the formulæ (35) come under this head, for in their case such a bilinear invariant is forthcoming in the determinant $(\lambda_2\mu_1 - \lambda_1\mu_2)$.

magnitudes which we can reduce to zero by many different assumptions of R_0, R_1, R_2. We can then, however, as we know, determine $\mathbf{B}_0$, $\mathbf{B}_1$, $\mathbf{B}_2$ directly by means of an icosahedral equation. This being done, we again apply the formulæ (39) or (38), except that we interchange the letters **A** and **B**, and therefore express $\mathbf{A}_0$, $\mathbf{A}_1$, $\mathbf{A}_2$ in terms of $\mathbf{B}_0$, $\mathbf{B}_1$, $\mathbf{B}_2$. The coefficients R_1, R_2, R_3 are then necessarily rational functions of the original A, B, C, D, and those irrationalities which we may have introduced in making $\mathbf{B}_0^2+\mathbf{B}_1\mathbf{B}_2=0$; *the original problem of the* **A**'*s is therefore solved through the intervention of these irrationalities and the icosahedral equation appropriate to the* **B**'*s.**

I have only explained the general process in order to allow the applicability of formula (39) to come to light. The course which we will now pursue in order to solve the problem of the **A**'s, *i.e.* to reduce it to an icosahedral equation, is a much simpler one. We had:

$$(49) \qquad 2\mathbf{A}_0 = -(\lambda_1\lambda_2' + \lambda_2\lambda_1'),\ \mathbf{A}_1 = \lambda_2\lambda_2',\ \mathbf{A}_2 = -\lambda_1\lambda_1';$$

we will now attempt the solution by supposing the icosahedral equation constructed on which depends the $\frac{\lambda_1}{\lambda_2}$ *or* $\frac{\lambda'_1}{\lambda'_2}$ *respectively, which here occurs.* Geometrically speaking, this means that we seek to determine the point **A** by means of one of the two points on $A=0$ in which a tangent from **A** to the conic A meets it, while the general method just sketched—though here we suppose the functions in question as homogeneous functions of $\mathbf{A}_0$, $\mathbf{A}_1$, $\mathbf{A}_2$—co-ordinates to the point **A** *any one* covariant point lying on $A=0$, and then considers its co-ordinates $\mathbf{B}_0$, $\mathbf{B}_1$, $\mathbf{B}_2$, determined not merely relatively, but absolutely.

The analogy of our statement of the question with that which we have dealt with in the preceding chapter, according to *Herr Gordan's* plan, is obvious. In both cases we are concerned, as we know, with a form-problem of which the variables are bilinear forms of two series of binary variables which are simultaneously subjected to the icosahedral substitutions; in both we seek the solution by returning to the icosahedral equation on which the variables of the one series (in so far as their ratios

* We have already mooted the same point (when speaking of the Jacobian equations of the sixth degree) in II, 1, § 6.

are concerned) depend. We shall accordingly be able to follow precisely the course of ideas which was developed in §§ 6–11 of the preceding chapter; the individual steps are so simple that it appears superfluous to build up in detail the several results.

We begin with enumerating these homogeneous integral functions of λ_1, λ_2, and λ_1', λ_2', which remain unaltered for the simultaneous (here cogredient) icosahedral substitutions of these magnitudes (*invariant forms*). We have placed side by side in formula (35) the two simplest forms linear in the λ''s; they were the following two:

$$(50)\qquad \begin{cases} \lambda_1\lambda_2' - \lambda_2\lambda_1' = \sqrt{A}, \\ \dfrac{\delta f}{\delta\lambda_1} \cdot \lambda_1' + \dfrac{\delta f}{\delta\lambda_2} \cdot \lambda_2' = \mathsf{P}, \end{cases}$$

(where the computed value of the first form is immediately assigned, and the letter P is henceforth introduced for the sake of brevity). To these belong further, as we remarked in § 2, all other forms which arise from $f(\lambda_1, \lambda_2)$, $H(\lambda_1, \lambda_2)$, $T(\lambda_1, \lambda_2)$ by polarisation with respect to λ_1', λ_2'.* Our A, B, C, D, the "known" magnitudes of the form-problems, are those combinations of the forms here mentioned which are symmetrical in the λ, λ''s.

We consider now generally the *interchange* of the λ, λ''s, *i.e.*, the replacement of λ_1, λ_2 by λ_1', λ_2', and conversely. If an invariant form remains unaltered for the interchange of λ, λ', it is an integral function of A, B, C, D; if, on the other hand, it changes its sign on permutation, it is the product of $\sqrt{A}$ (50) and such an integral function. If an invariant is only of the same degree in the λ, λ''s, it can always be put into the following form:

$$(51)\quad F(\lambda_1, \lambda_2;\ \lambda_1', \lambda_2') = G(A, B, C, D) + \sqrt{A} \cdot H(A, B, C, D),$$

where the integral functions G, H, are defined by means of the following equations:

$$(52)\qquad \begin{cases} 2G = F(\lambda_1, \lambda_2;\ \lambda_1', \lambda_2') + F(\lambda_1', \lambda_2';\ \lambda_1, \lambda_2), \\ 2\sqrt{A} \cdot H = F(\lambda_1, \lambda_2;\ \lambda_1', \lambda_2') - F(\lambda_1', \lambda_2';\ \lambda_1, \lambda_2).— \end{cases}$$

The general course of our method of solution will now be as follows. We have first to construct the icosahedral equation:

* I do not further press the point that with the forms thus enumerated the entire system of the invariants here coming under consideration is exhausted.

$$(53) \qquad \frac{H^3(\lambda_1, \lambda_2)}{1728 f^5(\lambda_1, \lambda_2)} = Z,$$

on which $\lambda_1 : \lambda_2$ depends, and then to express the invariant $\mathbf{P}$ (50) in terms of λ_1, λ_2, $\sqrt{A}$, and the known magnitudes. Both steps are accomplished by appropriate applications of formulæ (51), (52). We then consider the formulæ (50) as linear equations for the determination of λ_1', λ_2': *the final formulæ for* $\mathbf{A}_0$, $\mathbf{A}_1$, $\mathbf{A}_2$, *which we sought are given on introducing in* (49) *the values which are found.* Here $\mathbf{A}_0$, $\mathbf{A}_1$, $\mathbf{A}_2$, necessarily appear as particular linear combinations of the linear invariants $\sqrt{A}$ and $\mathbf{P}$.

§ 12. Corresponding Formulæ.

The formulæ which are required in virtue of the general method just given are now to be developed so far as appears desirable for giving preciseness to the course of our ideas. I will here again (as in the preceding chapter) denote the forms originally given us by the index 1, the others, which arise from them by interchange of the variables λ, λ', by the index 2. Higher indices may denote the degree of integral functions of the arguments adopted in each case, on the understanding that these arguments are considered as functions of $\mathbf{A}_0$, $\mathbf{A}_1$, $\mathbf{A}_2$.

We begin with the computation of Z (53), or, as we now say, of Z_1. We have evidently, in order:

$$(54) \qquad \begin{aligned} Z_1 &= \frac{H_1^3}{1728 f_1^5} = \frac{H_1^3 f_2^5}{1728 f_1^5 f_2^5} \\ &= \frac{(H_1^3 f_2^5 + H_2^3 f_1^5) + (H_1^3 f_2^5 - H_2^3 f_1^5)}{3456 f_1^5 f_2^5} \\ &= \frac{G_{60}(A, B, C) + \sqrt{A} \, . \, D \, . \, G_{44}(A, B, C)}{3456\, [G_{12}(A, B, C)]^5}. \end{aligned}$$

Besides (51) and (52), I have here made use of the fact that, among the given magnitudes A, B, C, D, only D is of uneven degree in the $\mathbf{A}$'s, and also that D^2 is an integral function of A, B, C. The integral functions G_{12}, G_{44}, G_{60} of A, B, C remain to be estimated by recurring to the explicit values of the magnitudes in $\mathbf{A}_0$, $\mathbf{A}_1$, $\mathbf{A}_2$, which come under consideration. The computation in question is, of course, somewhat formidable; I omit it, because it furnishes nothing of special interest.

We now turn to the computation of P, or rather of P_1. The form P_1 is of the first dimension in the λ''s, of the eleventh in the λ's; if we wish to employ a process like the one applied to Z_1, we shall have to first affect P_1 with such factors dependent only on λ_1, λ_2, that the aggregate which arises is uniformly of the first dimension in the λ, λ''s. We put accordingly:

$$\rho_1 = \frac{\mathsf{P}_1 \cdot H_1}{T_1}, \tag{55}$$

and then have, in order:

$$\begin{aligned}\rho_1 &= \frac{\mathsf{P}_1 \cdot H_1}{T_1} = \frac{\mathsf{P}_1 \cdot H_1 T_2}{T_1 T_2} \\ &= \frac{(\mathsf{P}_1 H_1 T_2 + \mathsf{P}_2 H_2 T_1) + (\mathsf{P}_1 H_1 T_2 - \mathsf{P}_2 H_2 T_1)}{2 T_1 T_2} \\ &= \frac{D \cdot G_{16}(A, B, C) + \sqrt{A} \cdot G_{30}(A, B, C)}{2\Gamma_{30}(A, B, C)},\end{aligned} \tag{56}$$

where the integral functions G_{16}, G_{30}, Γ_{30} remain to be evaluated.

On substituting, we have:

$$\mathsf{P}_1 = \frac{T_1}{H_1} \cdot \frac{D.G_{16}(A, B, C) + \sqrt{A} \cdot G_{30}(A, B, C)}{2\Gamma_{30}(A, B, C)}. \tag{57}$$

We now seek, as we suggested, to obtain the λ_1', λ_2' from $\sqrt{A}$ and P_1. The formulæ which arise run simply:

$$\begin{cases} \lambda_1' = -\sqrt{A} \cdot \dfrac{\frac{\delta f_1}{\delta \lambda_2}}{12 f_1} + \mathsf{P}_1 \cdot \dfrac{\lambda_1}{12 f_1}, \\[2ex] \lambda_2' = +\sqrt{A} \cdot \dfrac{\frac{\delta f_1}{\delta \lambda_1}}{12 f_1} + \mathsf{P}_1 \cdot \dfrac{\lambda_2}{12 f_1}. \end{cases} \tag{58}$$

Comparing it with (49) we have finally:

$$\begin{cases} 2\mathsf{A}_0 = -\sqrt{A} - 2\mathsf{P}_1 \cdot \dfrac{\lambda_1 \lambda_2}{12 f_1}, \\[2ex] \mathsf{A}_1 = +\sqrt{A} \cdot \dfrac{\lambda_2 \frac{\delta f_1}{\delta \lambda_1}}{12 f_1} + \mathsf{P}_1 \cdot \dfrac{\lambda_2^2}{12 f_1}, \\[2ex] \mathsf{A}_2 = -\sqrt{A} \cdot \dfrac{\lambda_1 \frac{\delta f_1}{\delta \lambda_2}}{12 f_1} - \mathsf{P}_1 \cdot \dfrac{\lambda_1^2}{12 f_1}, \end{cases} \tag{59}$$

where we suppose for P_1 the value (57) introduced.

We can in many respects modify the method of solution thus given if we like to take up once more the course of development adopted in the preceding chapter. Substitute, for example, in (59), instead of $\mathbf{P}_1$ the magnitude ρ_1 (55), so that the **A**'s depend only on $\sqrt{A}$, ρ_1, and $\lambda_1 : \lambda_2$, then compute the corresponding problem of the **A**'s regarding these three magnitudes as arbitrarily given, and compare it with the proposed problem. We thus obtain for ρ_1 and Z_1 (53) determining equations which can be applied to the actual computation of them. We can also, as we did in the preceding chapter, interpret geometrically each step of the method of solution. Leaving all these things to the reader, I emphasise, in conclusion, the appearance of $\sqrt{A}$. In the sense of our previous mode of expression, this is an *accessory irrationality*, *i.e.*, one which is not rational in the magnitudes $\mathbf{A}_0$, $\mathbf{A}_1$, $\mathbf{A}_2$ which are to be computed.* We shall soon see that an irrationality of this kind is, in fact, indispensable if we want to reduce the problem of the **A**'s to an icosahedral equation.

* In an analogous sense, the notion of the accessory irrationality is transferred generally to form-problems.

CHAPTER V

THE GENERAL EQUATION OF THE FIFTH DEGREE

§ 1. Formulation of Two Methods of Solution.

Turning now to the general equation of the fifth degree, let us proceed with the actual problem of solution.* We are in principle concerned with the construction, from five magnitudes $x_0, x_1, \ldots x_4$, which are subject to the single condition $\Sigma x = 0$, of a function $\phi(x_0, x_1, \ldots x_4) = \lambda$, which undergoes icosahedral substitutions for the even permutations of the x's. How we shall afterwards represent the individual x's in terms of λ is a question in itself which at first we regard as a secondary one. Limiting ourselves first to the main question, let us take a geometrical interpretation as our basis; we regard $x_0 : x_1 : \ldots x_4$, as we did above, as the co-ordinates of a point in space, λ as the parameter of a generator of the first kind on the canonical surface of the second degree $\Sigma x^2 = 0$. Our problem then becomes: *to any point x in space to co-ordinate by appropriate construction a generator λ in a covariant manner*, *i.e.*, to co-ordinate in such wise that the relation between the point and the generator remains unaltered when both are simultaneously subjected to the even collineations.

A first solution of this problem arises of its own accord, so to say, on the ground of the developments already given. *Namely, we shall at the outset exhibit in covariant relation to the point x a point y of the canonical surface, and then take as generator λ the generator passing through y.* Therefore, to characterise at once the algebraical treatment of the general equation of the fifth degree which arises from this, *we shall transform the*

* The developments given in the following pages are contained in their general features in my oft-cited works in Bd. 12, 14, 15 of the Mathematische Annalen, but they are here for the first time expounded in a connected form.

general equation of the fifth degree by an appropriate Tschirnhausian transformation into a canonical equation, and then solve this in accordance with the method expounded in the third chapter of the present part.

The Tschirnhausian transformation which is required in the process now described has been mentioned in greater detail in II, 2, § 6, and there formulated in the following manner: we first co-ordinated to the point x a straight line in space which joins two rational points covariant to x, and then chose as our point y one of the points of intersection which this straight line has in common with the canonical surface. Here, generally speaking, an accessory square root would be necessary for separating the two points of intersection. If we wish to express ourselves briefly, we can even put aside the point y in our description of this construction. Our object is then simply to employ one of the two generators of the first kind on the canonical surface, which meet a straight line which is covariant to x. The accessory square root depends on the fact that alongside of a first generator of this kind which we call λ, there is always a second associated with λ, which we will denote by λ' for a moment. Expressing ourselves thus, we recognise the possibility of still further postponing the use of the accessory irrationality. *Instead of seeking at once the icosahedral equation on which λ depends, we shall first establish the equation system by which the symmetric functions of λ, λ' are determined, and not till later deduce from this equation system the aforesaid icosahedral equation.* But this is manifestly the same as saying that we return to the developments of the fourth chapter which we have just concluded. In fact, our λ, λ' are cogredient variables; the equation system of which we speak is therefore an equation system of the **A**'s, in the treatment of which we shall, moreover, be led at once, as we shall see, to the homogeneous arrangement, *i.e.*, to the *form-problem of the* **A***'s*. At the same time, the icosahedral equation on which λ depends is the same as we should anyhow use in the solution of the problem of the **A**'s. *We therefore find a second method of solution of the general equation of the fifth degree, in which we turn to account the developments of* II, 4, *exactly in the same way as we do those of* II, 3 *for the first method of solution.*

For the rest, the formulation which we have just established for the second method of solution is unnecessarily precise. Recalling the considerations which we have given in II, 2, § 9, we recognise that we can co-ordinate to the point x, instead of a straight line, a general linear complex in order to effect the second method. *The generators* λ, λ' *are then those two which belong to this linear complex.* The explicit formulæ, which we shall establish later with a view to giving exactness to the second method, remain undisturbed by this generalisation; we shall therefore only quite cursorily return to the special formulation which we just now began.

We have now a twofold task in the following paragraphs. In the first place, we shall have to establish in detail the formulæ which correspond to the two methods of solution, the feasibility of which we ascertained; and then we will bring the totality of those researches which we summarised in II, 1 into conformity with our own reflexions. In this respect the relationship of our first method of solution with that of *Bring* and of our second method with that of *Kronecker* is evident at the outset. By using a theorem which we previously established (I, 2, § 8) concerning the icosahedral substitutions, we then succeed in proving also that fundamental proposition of the Kronecker theory to which we referred in II, 1, § 7.

§ 2. Accomplishment of our First Method.

In order to give exactness to our first method, let

(1) $$x^5 + ax^3 + bx^2 + cx + d = 0$$

be the given equation of the fifth degree (in which we have taken $\Sigma(x) = 0$). We then further put, in accordance with II, 2, § 5.

(2) $$y_\nu = p \cdot x_\nu^{(1)} + q \cdot x_\nu^{(2)} + r \cdot x_\nu^{(3)} + s \cdot x_\nu^{(4)},$$

where $x_\nu^{(k)} = x_\nu^k - \frac{1}{5}\Sigma x^k$, and compute Σy^2. This is a homogeneous integral function of the second degree of p, q, r, s:

(3) $$\Phi\ (p, q, r, s),$$

of which the coefficients are symmetrical integral functions of the x's, and therefore integral functions of the coefficients a, b, c, d occurring in (1). *We wish to find a solution system of* *the equation $\Phi=0$ which remains unaltered for the even permutations of the x's.*

Let us first remark that the p, q, r, s required cannot possibly be equal to rational functions of $x_0, x_1, \ldots x_4$. This follows from the proof, to be presently effected, that the use of an accessory irrationality, at least therefore an accessory square root, is indispensable for carrying out our method.* We return the more readily to the geometrical construction with the *covariant straight lines* which we described just now, and for which we have given the necessary formulæ in II, 2, § 6. Let:

$$P_1, Q_1, R_1, S_1; \; P_2, Q_2, R_2, S_2$$

be two series of four magnitudes which depend rationally on the x's, and in such wise that they are not altered for the even permutations of the x's, and which are therefore rational functions of the coefficients a, b, c, d of (1) and of the square root of the corresponding discriminant.

We then put in (2) as before:

$$(4) \quad p=\rho_1 P_1+\rho_2 P_2, \; q=\rho_1 Q_1+\rho_2 Q_2, \; r=\rho_1 R_1+\rho_2 R_2, \; s=\rho_1 S_1+\rho_2 S_2.$$

By this means Φ [formula (3)] is transformed into a binary quadratic form of the ρ_1, ρ_2's, the coefficients of which are rational functions of the known magnitudes: we put $\Phi=0$, and determine ρ_1, ρ_2, from the quadratic equation which arises, whereby the proposed accessory square root is introduced. Then let us substitute corresponding values of ρ_1, ρ_2, in (4) and (2) respectively, and compute the canonical equation which results for the y's, which we will briefly denote as follows:

$$(5) \qquad y^5+5\alpha y^2+6\beta y+\gamma=0.$$

Thus we have made every arrangement necessary for the imme-

* Conversely, if we proved the theorem in the text (concerning the irrationality of p, q, r, s) directly, we should have a new proof of the necessity of the square root in question. Namely, could an icosahedral equation be produced from (1) without employing accessory irrationalities, we should be able to construct from this one of the infinitely many corresponding canonical resolvents of the fifth degree, and then obtain, by collecting the formulæ, a transformation (2) of which the coefficients p, q, r, s would be rational functions of the x's, unaltered by the even permutations of the x's.

diate application of the developments of II, 3. If we have then computed the roots y_ν of (5) by the help of these developments, we find the corresponding x_ν's by reversing (2).

I should like here to make a passing remark with regard to the inversion of the Tschirnhausian transformation. It is usually said, and we have also so expressed it farther back (II, 1, § 1), that the x_ν required is computed rationally as a common root of the equations (1) and (2) [where now y_ν is to be regarded as the unknown magnitude]. It is essentially the same, but more in the spirit of the rest of our considerations, if we place opposite formula (2) another explicit one:

$$(6) \qquad x_\nu = p' \cdot y_\nu^{(1)} + q' \cdot y_\nu^{(2)} + r' \cdot y_\nu^{(3)} + s' \cdot y_\nu^{(4)},$$

where $y_\nu^{(k)} = y_\nu^k - \frac{1}{5}\Sigma y^k$ and p', q', r', s' denote rational functions of ρ_1, ρ_2, a, b, c, d, and of the square root of the discriminant of (1), which are computed by the aid of elementary methods. The determination of the x_ν's can then be conceived in this sense, that, geometrically speaking, we derive from the first found point, $y = y^{(1)}$, three more covariant points, $y^{(2)}$, $y^{(3)}$, $y^{(4)}$, and then construct the required point x by means of invariant coefficients.*

Let us consider that in the method of solution here explained the computation of x_ν from the root λ of the icosahedral equation finally established is divided into two steps; we have originally, in II, 3, employed the five-valued functions of λ:

$$u_\nu = \frac{12 f^2(\lambda) \cdot t_\nu(\lambda)}{T(\lambda)}, \quad v_\nu = \frac{12 f(\lambda) \cdot W_\nu(\lambda)}{H(\lambda)}$$

in order to compose the y_ν's linearly from them, or from v_ν and $u_\nu v_\nu$; we have then represented the point x as a linear combination of $y^{(1)}$, $y^{(2)}$, $y^{(3)}$, $y^{(4)}$. We can evidently condense these two steps into one: *we can construct the point x from four points which are covariants of the generator λ.* The simplest rational functions of λ, which assume on the whole four values for the

* The geometrical mode of expression in the text is of course only a counterpart of the algebraical process, when the latter is so specialised that the law of homogeneity is satisfied throughout, *i.e.*, that the ratios of the y's only depend on the ratios of the x's. We ought really to repeat the same remark in all the following developments, but for the sake of brevity we shall omit doing so.

icosahedral substitutions, are, by what precedes (I, 4), the following:

$$u_\nu,\ v_\nu,\ u_\nu v_\nu,\ r_\nu = \frac{t^2(\lambda)}{f(\lambda)}.$$

Here $\Sigma u = \Sigma v = \Sigma uv = 0$, while on the other hand $\Sigma r \lessgtr 0$, so that instead of r_ν we will introduce the combination $r_\nu - \frac{1}{5}\Sigma r$. Then:

$$(7) \qquad x_\nu = p'' \cdot u_\nu + q'' \cdot v_\nu + r'' \cdot u_\nu v_\nu + s'' \cdot \left(r_\nu - \frac{1}{5}\Sigma r\right),$$

where p'', q'', r'', s'' are coefficients of the same nature as p', q', r', s'.

I have appended this new formula of inversion really for the sake of completeness. In fact, it is just this which appears to me to be the peculiar advantage of our first method—that when formulated in the way represented by (6) it is decomposed into two separate parts, of which the first, which is concerned with the connection of the general equation of the fifth degree with the canonical equation, has throughout quite an elementary character. We can also consider formula (6) as more simple than formula (7). Namely, if we consider P_1, Q_1, . . . R_2, S_2 in (4) as rationally dependent on a, b, c, d alone, not on the square root of the corresponding discriminant, then the square root of the discriminant will also be wanting in the coefficients of (6), while it necessarily appears in the coefficients of (7), as also in the right-hand side of the icosahedral equation for λ.

§ 3. Criticism of the Methods of Bring and Hermite.

Before going further, we shall compare our first method of solution with the closely related kinds of solution which Bring and Hermite have respectively given. The details which here come under consideration have already been developed in II, 3, §§ 13, 14. *Now that we revert to these, we must describe our method as an essential simplification of the Bring method.* Bring, as we do, transforms the given equation of the fifth degree into a canonical equation; he, too, employs the rectilinear generators which lie on the canonical surface. But beyond this he comes to an unnecessary complication: in order to obtain a normal equation with only one parameter, he thinks that a new accessory irrationality has to be introduced by the intervention of an auxiliary equation of the third degree. I, on my part, maintain that the original process of Bring should be given up, and replaced by our first method, *which retains the essential idea of the Bring method.* The advance with which we are concerned finds significant expression in the "deficiency" of the figure to be employed for the geometrical interpretation; the family of rectilinear generators (lying on the canonical surface) of the one kind or the other form a manifoldness of deficiency $p=0$; the deficiency of the Bring curve is equal to 4.*

As the crowning point we shall embrace in the critique thus formulated the process of *Hermite* also: *if we wish to apply elliptic functions to the solution of the canonical equation of the fifth degree, this is done most simply by using the formula given in* I, 5, § 7, *for the root of the corresponding icosahedral equation.*

Hermite's use of the Bring form can only come under consideration thenceforward if, instead of the rational invariant $\frac{g_2^3}{\Delta}$, to which the right-hand side of the icosahedral equation is equal, we employ the corresponding κ^2. In fact, we saw in II, 3, § 14, that the cubic auxiliary equation of Bring becomes reducible when we consider κ as known. I will also here bring into special prominence the advance which is made by our having deduced directly from the *form* of the icosahedral equation the possibility of solving the icosahedral equation by means of elliptic functions. (*Vide* I, 5, § 7.)

§ 4. Preparation for our Second Method of Solution.

The geometrical opening which we have given for our second method of solution requires us to establish the quadratic equation on which depend the two generators of the first kind on the canonical surface which belong to a definite linear complex. We have solved this very problem in II, 2, § 10, for any

* Starting from the value of p and the general theory of curves with $p=4$, we can show (as I cannot do here in detail) that Bring's cubic auxiliary equation is, in fact, indispensable if we would determine a point of the Bring curve, *i.e.*, employ the trinomial equation $y^5+5\beta y+\gamma=0$ as normal equation.

surface of the second degree, *though only for the case, we must admit, of a particular system of co-ordinates.* We had then taken as the equation of the surface the following:

$$(8) \qquad X_1X_4 + X_2X_3 = 0,$$

and had then defined the parameter λ of the generator of the first kind in the following manner:

$$(9) \qquad \lambda = -\frac{X_1}{X_2} = \frac{X_3}{X_4} = \frac{\lambda_1}{\lambda_2},$$

and, finally, understanding by $A_{\mu,\nu}$, the co-ordinates of the linear complex, we had obtained the equation:

$$(10) \qquad A_{42}\lambda_1^2 + (A_{23} - A_{14})\lambda_1\lambda_2 + A_{13}\lambda_2^2 = 0.$$

I add here at once the corresponding formulæ for the generator of the second kind. We found as the defining equation of the parameter μ:

$$(11) \qquad \mu = -\frac{X_2}{X_4} = \frac{X_3}{X_1} = \frac{\mu_1}{\mu_2},$$

and as the corresponding quadratic equation:

$$(12) \qquad -A_{34}\mu_1^2 + (A_{23} + A_{14})\mu_1\mu_2 + A_{12}\mu_2^2 = 0.$$

We now recall the method by which we introduced the parameters λ, μ for the canonical surface in particular, in II, 3, §§ 2, 3. This was done in exact agreement with (8), (9), (11), only that instead of X_1, X_2, X_3, X_4, we wrote p_1, p_2, p_3, p_4 respectively, where p_μ denoted the expression of Lagrange:

$$(13) \qquad p_\mu = x_0 + \epsilon^\mu \, . \, x_1 + \epsilon^{2\mu} \, . \, x_2 + \epsilon^{3\mu} \, . \, x_3 + \epsilon^{4\mu} \, . \, x_4.$$

We can therefore retain equations (10), (12), *unaltered, provided we proceed throughout on the basis of Lagrange's system of co-ordinates in dealing with them.*

In II, 2, § 9, where we co-ordinated to the point x a co-variant linear complex in the most general manner, this latter supposition has not been made; the co-ordinates there given for the complex:

$$(14) \qquad a_{ik} = \sum_{l,\,m} c^{l,\,m}\left\{x_i^{(l)}x_k^{(m)} - x_i^{(m)}x_k^{(l)}\right\}$$

[where the $c^{l,m}$ denote any rational functions of the coefficients a, b, c, d and the square root of the corresponding discriminant] are related to the fundamental pentahedron, as are also the point-coördinates x themselves. Our first task is therefore a transformation of co-ordinates: *we must determine the co-ordinates* $A_{\mu\nu}$ *which the complex* (14) *assumes if we introduce, by means of* (13), *the expressions* p_μ. To this end I will denote those p's which belong to the points $x^{(l)}$, $x^{(m)}$, by $p^{(l)}$, $p^{(m)}$. We then have:

$$(15)\qquad p_\mu^{(l)}p_\nu^{(m)} - p_\nu^{(l)}p_\mu^{(m)} = \sum_{i,k}(\epsilon^{\mu i+\nu k} - \epsilon^{\nu i+\mu k})\,(x_i^{(l)}x_k^{(m)} - x_k^{(l)}x_i^{(m)}),$$

where on the right-hand side each combination $(i, k)=(k, i)$ occurs once. We now add the six equations which we obtain in this manner for the different combinations $(l, m)=(m, l)$, after we have multiplied each by $c^{l,m}$ [formula (14)]. There thus remain on the left side the $A_{\mu\nu}$'s required, while on the right side sets of six terms are condensed into the a_{ik}'s (14). *Hence the formulæ of transformation which we seek run thus:*

$$(16)\qquad A_{\mu\nu} = \sum_{i,k}(\epsilon^{\mu i+\nu k} - \epsilon^{\nu i+\mu k})\,.\,a_{ik}.$$

We now introduce the $A_{\mu\nu}$'s so obtained into (10), (12). I will write the quadratic equations which here arise in the form which we just now established in II, 4:

$$(17)\qquad \begin{cases} \mathsf{A}_1\lambda_1^2 + 2\mathsf{A}_0\lambda_1\lambda_2 - \mathsf{A}_2\lambda_2^2 = 0 \\ \mathsf{A}_1'\mu_1^2 + 2\mathsf{A}_0'\mu_1\mu_2 - \mathsf{A}_2'\mu_2^2 = 0. \end{cases}$$

We then have:

$$(18)\qquad \begin{cases} 2\mathsf{A}_0 = +A_{23} - A_{14} = \sum_{i,k}(\epsilon^{2i+3k} - \epsilon^{3i+2k} + \epsilon^{4i+k} - \epsilon^{i+4k})\,.\,a_{ik}, \\ \mathsf{A}_1 = +A_{42} \qquad = \sum_{i,k}(\epsilon^{4i+2k} - \epsilon^{2i+4k})\,.\,a_{ik}, \\ \mathsf{A}_2 = -A_{13} \qquad = \sum_{i,k}(\epsilon^{3i+k} - \epsilon^{i+3k})\,.\,a_{ik}, \end{cases}$$

likewise also;

$$(19)\begin{cases} 2\mathsf{A}_0' = -A_{23} - A_{14} = \sum_{i,k}(\epsilon^{3i+2k} - \epsilon^{2i+3k} + \epsilon^{4i+k} - \epsilon^{i+4k}) \,.\, a_{ik}, \\ \mathsf{A}_1' = +A_{34} = \sum_{i,k}(\epsilon^{3i+4k} - \epsilon^{4i+3k}) \,.\, a_{ik}, \\ \mathsf{A}_2' = +A_{12} = \sum_{i,k}(\epsilon^{i+2k} - \epsilon^{2i+k}) \,.\, a_{ik}. \end{cases}$$

§ 5. Of the Substitutions of the A, A′'s—Definite Formulation.

In virtue of the geometrical considerations which we have placed in the foreground, it is manifest that the *ratios* of the $\mathsf{A}_0, \mathsf{A}_1, \mathsf{A}_2$ (18) just established, undergo exactly the same linear substitutions for the even permutations of x_0, x_1, x_2, x_3, x_4 as the *ratios* of magnitudes established in the preceding chapter, and denoted by the same letters; it is likewise evident that the ratios of the A′'s introduced in (19) behave contragrediently to the ratios of the A's. I say now that this correspondence holds good if we regard, instead of the ratios of the A, A′'s, the A, A′'s themselves. It would not be hard to prove the accuracy of this assertion on general grounds. We shall presently, § 9, indicate the method of doing so; meanwhile let us be satisfied with verifying its accuracy from the formulæ. We have evidently only to take the two operations S, T into consideration, all the others being composed from these by iteration and combination. First, as regards the even permutations of the x's, we have in II, 3, § 2, introduced for S, T the following:

$$(20)\quad\begin{cases} S: x_\nu' = x_{\nu+1}, \\ T: x_0' = x_0,\ x_1' = x_2,\ x_2' = x_1,\ x_3' = x_4,\ x_4' = x_3. \end{cases}$$

Corresponding to them, we obtain definite permutations of the a_{ik}'s (14), and, if we take account of these, we have the following substitutions for the A's defined by (18):

$$(21)\quad\begin{cases} S: \quad \mathsf{A}_0' = \mathsf{A}_0, \quad \mathsf{A}_1' = \epsilon^4\mathsf{A}_1, \quad \mathsf{A}_2' = \epsilon\mathsf{A}_2; \\ T: \begin{cases} \sqrt{5}\,.\,\mathsf{A}_0' = \mathsf{A}_0 + \mathsf{A}_1 + \mathsf{A}_2, \\ \sqrt{5}\,.\,\mathsf{A}_1' = 2\mathsf{A}_0 + (\epsilon^2+\epsilon^3)\mathsf{A}_1 + (\epsilon+\epsilon^4)\mathsf{A}_2, \\ \sqrt{5}\,.\,\mathsf{A}_2' = 2\mathsf{A}_0 + (\epsilon+\epsilon^4)\mathsf{A}_1 + (\epsilon^2+\epsilon^3)\mathsf{A}_2, \end{cases} \end{cases}$$

i.e., exactly the same substitutions as we have given in II, 4,

§ 2.* As regards, however, the contragredience of the magnitudes A' (19) and the A's (18), it is sufficient to remark that the values of the A''s proceed from those of the A's if in the latter we change ϵ into ϵ^2 throughout.

We now suppose any of the invariant forms of the A, A''s (18), (19) constructed, such as we described in the preceding chapter; either therefore the expressions A, B, C, D, from the A's alone, or from the A, A''s simultaneously the functions F_1, F_2, F_3, linear in the A''s, which we have considered in § 9 of the same chapter [see especially formulæ (45), (46), (47).] On introducing for the A, A''s the corresponding values in $x_0, x_1, \ldots x_4$, we obtain throughout rational functions of the x's such as do not alter for the even permutations of the x's, and which therefore admit of expression, by the help of elementary methods which we do not carry out, as rational functions of the coefficients a, b, c, d occurring in (1), and of the square root of the corresponding discriminant. In order to formulate our second method in a definite manner, we at first employ only the problem of the A's, and therefore the values of the magnitudes just mentioned, A, B, C, D. We then follow the developments which we have given in the two concluding paragraphs of the preceding chapter, and construct, after adjoining the accessory irrationality $\sqrt{A}$, a corresponding icosahedral equation for the determination of λ. The only question which remains is how we will conversely express the roots $x_0, x_1, \ldots x_4$ by the help of this λ. This is to be dealt with in the following paragraph.

§ 6. The Formulæ of Inversion of our Second Method.

In order to solve the problem which still remains, no less than three methods present themselves, viz., according as we wish to solve our problem at one stroke or resolve it into two or three steps.

In the former case we make immediate use of the formula

* The letters A_0', A_1', A_2' are employed in (21) in quite a different sense from that of (19); as I do not recur hereafter to (21), no misunderstanding will, I hope, arise from this.

(7), which I again exhibit (laying aside now the accents there employed for p, q, r, s):

$$(22) \qquad x_\nu = p \,.\, u_\nu + q \,.\, v_\nu + r \,.\, u_\nu v_\nu + s\left(r_\nu - \frac{1}{5}\Sigma r^2\right).$$

Here p, q, r, s are rational functions of a, b, c, d of the square root of the corresponding discriminant, and the accessory square root $\sqrt{A}$.

In the second case we first express $\mathsf{A}_0, \mathsf{A}_1, \mathsf{A}_2$ as we did in detail in § 12 of the preceding chapter, in terms of the root λ of the icosahedral equation. We then further bring to our aid the lowest five-valued integral functions of the A's. According to § 5 of the preceding chapter, these are:

$$\delta_\nu, \; \delta_\nu', \; \delta_\nu^2, \; \delta_\nu\delta_\nu',$$

Here again $\Sigma\delta = \Sigma\delta' = \Sigma\delta\delta' = 0$, while $\Sigma\delta^2$ is different from zero, so that for the expression of the x_ν's we will introduce, instead of the individual δ_ν^2's, the combination $(\delta_\nu^2 - \frac{1}{5}\Sigma\delta^2)$. We have then again formulæ of the following kind:

$$(23) \qquad x_\nu = p' \,.\, \delta_\nu + q' \,.\, \delta_\nu' + r' . \left(\delta_\nu^2 - \frac{1}{5}\Sigma\delta^2\right) + s' \,.\, \delta_\nu\delta_\nu',$$

where p', q', r', s' are rational functions of a, b, c, d, and the square root of the discriminant, *but no longer contain the accessory square root* $\sqrt{A}$.

Finally, in the third case, we first suppose $\mathsf{A}_0, \mathsf{A}_1, \mathsf{A}_2$ again computed from the root λ of the icosahedral equation; but then, instead of seeking the x_ν's directly, we first seek the corresponding $\mathsf{A}_0', \mathsf{A}_1', \mathsf{A}_2'$ (19). We effect this by calculations analogous to those which we previously made on expressing the forms F_1, F_2, F_3 just mentioned, which depend on the A, A''s, as functions of a, b, c, d, and of the square root of the discriminant, and determining $\mathsf{A}_0', \mathsf{A}_1', \mathsf{A}_2'$ as unknowns occurring linearly. This being done, we seek the simplest possible functions of the A, A''s, which are five-valued and at the same time symmetric in the A, A''s. We find a first function of this kind if we square the y_ν of II, 3:

$$y_\nu = \epsilon^{4\nu} \,.\, \lambda_1\mu_1 - \epsilon^{3\nu} \,.\, \lambda_2\mu_1 + \epsilon^{2\nu} \,.\, \lambda_1\mu_2 + \epsilon_\nu \,.\, \lambda_2\mu_2,$$

and submit it to the process of transference continually employed in § 2 of the preceding chapter. In this manner there arises a form bilinear in the A, A''s:

$$(24) \qquad \chi_\nu = 2\mathsf{A}_0'(\epsilon^{4\nu}\mathsf{A}_1 + \epsilon^\nu\mathsf{A}_2) + \mathsf{A}_1'(-2\epsilon^{3\nu}\mathsf{A}_0 + \epsilon^{2\nu}\mathsf{A}_1 + \epsilon^{4\nu}\mathsf{A}_2) + \mathsf{A}_2'(-2\epsilon^{2\nu}\mathsf{A}_0 - \epsilon^\nu\mathsf{A}_1 + \epsilon^{3\nu}\mathsf{A}_2).$$

As other functions with the same properties, we will employ the powers χ_ν^2, χ_ν^3, χ_ν^4, where, however, we must consider that none of the power-sums $\Sigma\chi^2$, $\Sigma\chi^3$, $\Sigma\chi^4$ vanish identically. We shall therefore do best to write the formula which corresponds to (22), (23), with an extra term as follows:

$$(25) \qquad x_\nu = p'' \cdot \chi_\nu + q'' \cdot \chi_\nu^2 + r'' \cdot \chi_\nu^3 + s'' \cdot \chi_\nu^4 + t''.$$

Here p'', q'', r'', s'', t'' are again at the outset rational functions of a, b, c, d, and of the square root of the discriminant. *Moreover, we can arrange so that they shall be merely rational functions of a, b, c, d.* We have then merely to make the $c^{l,m}$'s in the original method (14) themselves depend rationally only on a, b, c, d.

I have brought together these data without detailed elaboration, because they, so to say, of necessity proceed from the previous developments. The third method of procedure appears to me unquestionably the most effective. Decomposing, as it does, the computation of the x_ν's into not less than three separate steps, it employs three times over the same elements of the typical exposition with which we have become acquainted under varying forms in the three preceding chapters.

§ 7. Relations to Kronecker and Brioschi.

Our second method of solution is, as we have often said before, only a *modification and extension* of the Kronecker method. In fact, we have seen in detail in II, 4 that the problem of the A's, in the sense there explained, can be replaced by its simplest resolvent of the sixth degree, the Jacobian equation. In the details many points of difference certainly present themselves. I will here only call attention to two, of which the second is the more important.

We first remark that the way in which Herr Kronecker,

in his first communication to Hermite,* reduces the general Jacobian equation of the sixth degree to the case $A=0$, or, as we now say, to an icosahedral equation, is different from the method applied in the preceding chapter. *Herr Kronecker so formulates his method that* A_0, A_1, A_2 *contain a parameter* ν *which occurs linearly, and which is afterwards so determined that* $\mathsf{A}_0^2+\mathsf{A}_1\mathsf{A}_2=A$ *becomes zero.* We can, of course, combine this idea with our formulæ, viz., by providing at the outset the $c^{l,m}$'s themselves [formula (14)] with a parameter λ occurring linearly. Then, instead of distinguishing by an accessory square root the two generators of the first kind, which the linear complex in question for any value of ν has in common with the canonical surface, we proceed thus: we first make the complex variable in a linear fasciculus, and then fix its position by the condition that it shall contain two coincident generators of the first kind belonging to the canonical surface. This condition itself brings with it an accessory square root. I have in what precedes dispensed with the formulation thus pointed out, because it is only applicable if we treat the problem of the A's as a resolvent of the proposed equation of the fifth degree, while I wished to first consider the problem of the A's independently of such connections.

We further remark *that the general formulæ which Signor Brioschi has given for the accomplishment of the Kronecker method,* formulæ of which we gave a detailed account in II, 1, § 6, *are throughout different from our formulæ* (18). Signor Brioschi employs for the construction of his A_0, A_1, A_2 six linearly independent magnitudes u_∞, u_0, . . . u_4, while we use twenty magnitudes a_{ik}, between which the relations $a_{ik}=-a_{ik}$, $\sum_i a_{ik}=\sum_k a_{ik}=0$ subsist. Again we are satisfied with the same magnitudes, a_{ik}, when we wish to take under consideration the A''s alongside of the A's, while Signor Brioschi would have to annex six new magnitudes u_∞', u_0', . . . u_4'. I will not pursue this comparison, which only concerns the *external configuration* of the formulæ, any further. Let us remark, above all, *that our formulæ* (quite as much as those of Brioschi) *are in any case as general as they can be.* If, namely, the A, A''s are arbitrarily given, we can from them determine conversely the correspond-

* See II, 1, § 6.

ing a_{ik}'s and $c^{l,m}$'s respectively [formula (14)]. We have only to repeat the transformation of co-ordinates of § 4 in a reverse sense.

The calculation in question takes the following form. We have first, on returning from (18), (19), to the co-ordinates, $A_{\mu\nu}$ (16):

$$(26) \qquad \begin{cases} A_{12} = \mathsf{A}_2', & A_{34} = \mathsf{A}_1', \\ A_{13} = -\mathsf{A}_2, & A_{42} = \mathsf{A}_1, \\ A_{14} = -\mathsf{A}_0 - \mathsf{A}_0', & A_{23} = \mathsf{A}_0 - \mathsf{A}_0'. \end{cases}$$

We then replace the formulæ (13) by their reciprocals

$$(27) \qquad 5x_i = \epsilon^{-i} \cdot p_1 + \epsilon^{-2i} \cdot p_2 + \epsilon^{-3i} \cdot p_3 + \epsilon^{-4i} \cdot p_4.$$

Hence:

$$25\,(x_i^{(l)}x_k^{(m)} - x_k^{(l)}x_i^{(m)}) = \sum_{\mu,\,\nu} (\epsilon^{-\mu i - \nu k} - \epsilon^{-\nu i - \mu k})\,(p_\mu^{(l)}p_\nu^{(m)} - p_\nu^{(l)}p_\mu^{(m)}),$$

where the summation on the right-hand extends over all combinations $(\mu, \nu) = (\nu, \mu)$, and now, on multiplying the individual equation by $c^{l,m}$ and adding the several terms for $(l, m) = (m, l)$:

$$(28) \qquad 25a_{ik} = \sum_{\mu,\,\nu} (\epsilon^{-\mu i - \nu k} - \epsilon^{-\nu i - \mu k}) \cdot A_{\mu\nu},$$

which is the formula we sought.

I should like, in conclusion, to formulate concisely once more the geometrical idea which lies at the root of our treatment of the Kronecker method, and which probably possesses far-reaching significance. The first thing is, that we substitute in general for the point x a linear complex, considering, therefore, instead of the equation of the fifth degree, an equation of the 20th degree whose roots a_{ik} satisfy the oft-mentioned relations $a_{ik} = -a_{ki}$, $\sum\limits_i a_{ik} = \sum\limits_k a_{ik} = 0$. The second thing is, that we refer this complex by means of (18), (19), to a new system of co-ordinates. I will not enter * into any details concerning the significance of the A, A''s, but only remark that the first

* Consult my essay in the second volume of the Annalen (1869): "Die allgemeine lineare Transformation der Liniencoordinaten." Consider, in particular, that the linear complex becomes a special one, *i.e.*, a straight line, when $\mathsf{A}_0^2 + \mathsf{A}_1\mathsf{A}_2 = \mathsf{A}_0'^2 + \mathsf{A}_1'\mathsf{A}_2'$.

of the two equations (17) vanishes identically when all the **A**'s are equal to zero, the second when all the **A**′'s are zero. *Therefore, for the generators of the first kind of the canonical surface,* $\mathbf{A}_0=\mathbf{A}_1=\mathbf{A}_2=0'$, *for the generators of the second kind,* $\mathbf{A}_0'=\mathbf{A}_1'=\mathbf{A}_2'=0$. What is the object of this transformation of co-ordinates? By its means we are enabled to replace the equation of the 20th degree for the a_{ik}'s by *the form-problem of the* **A**'*s or the* **A**′'*s.* In fact, we have seen that, for the 60 even collineations of space, $\mathbf{A}_0$, $\mathbf{A}_1$, $\mathbf{A}_2$, and likewise $\mathbf{A}_0'$, $\mathbf{A}_1'$, $\mathbf{A}_2'$, are linearly substituted in their own right, and therefore as ternary forms. Let us now remark that we could have premised, *a priori*, this property of our geometrical conception. Namely, for the even collineations of space each of the two systems of rectilinear generators of the canonical surface becomes transformed, as we know, into itself. *Hence of necessity the two threefold families of linear complexes to which these systems of generators respectively belong are also transformed into themselves for these collineations.* But from this follows directly the property of the **A**, **A**′'s described, provided we further postulate that to every collineation corresponds a linear transformation of the line co-ordinates. *The possibility of reducing our equation of the* a_{ik}*'s to a ternary form-problem thus appears as an immediate outcome of the elementary intuitions of line geometry.* This is the particular point of view under which I should like to see the second method considered.

§ 8. Comparison of our Two Methods.

The two methods for the solution of the equation of the fifth degree which we have contrasted with one another are, nevertheless, as follows from the considerations of § 1 of the present chapter, very intimately related: we will here show that it is only in non-essential points that they differ, inasmuch as every icosahedral equation which is co-ordinated to a proposed equation of the fifth degree by virtue of the one method can always be deduced by means of the other method.

The passage in this sense, from the first method to the second, is immediately evident. In order to co-ordinate a point y of the canonical surface as a covariant to the point x, we have just now (§ 2) constructed first a straight line covariant to x, and have then intersected with this the canonical surface. *We can now start from this very line as the special linear complex of the second method:* we have only to compute the corresponding co-ordinates a_{ik}. If we, then, construct the corresponding problem of the **A**'s, one of the two icosahedral equations by which we can solve this problem will immediately become identical with the icosahedral equation to which the determination of the y_ν's leads.

The converse of this argument is not much more complicated. We assume that we have by means of our second method co-ordinated to the equation of the fifth degree an icosahedral equation, and therefore to the point x a generator λ of the canonical surface. *Then we can always find in a rational manner* (and this in many different ways) *a point y which lies on the generator λ*: we need only, for example, make the y_ν's proportional to the $W_\nu(\lambda)$'s or to the other expressions which occur in the principal resolvent of the icosahedral equation. But this point is co-ordinated to the point x in any case as a covariant; *we have therefore at once a Tschirnhausian transformation which co-ordinates to the point x a point y on the canonical surface.* If we now make this Tschirnhausian transformation the basis of our second method, we return of course to the initial icosahedral equation.

In this sense we can say that in reality *only one solution of the equation of the fifth degree is found.* The difference between the two methods which we proposed *lies only in the order of the individual steps.* In the first method we give prominence to the accessory square root, in the second we do not introduce it till after separating the two systems of generators. Against this, the first method, as we have said before, has the advantage of operating at first with quite elementary material.

Howbeit the common foundation of the two methods in our exposition appears to be first the theory of the icosahedron, and then further the consideration of the rectilinear generators of the canonical surface. That the first gives the actual normal equations to which we must once for all reduce the solution of the equation of the fifth degree, I cannot doubt. On the other hand, I form a different estimate of our geometrical reflexions and constructions, useful as they have been to us. I believe that we shall be enabled to develop the general theory of form-problems algebraically, and in such wise that our reduction of equations of the fifth degree to the icosahedron appears as a mere corollary, and does not need to be established in a special manner. I have myself attempted this in Bd. xv of the Mathematische Annalen, where I brought the connection between the problem of the **A**'s and of the equation of the fifth degree—and, in fact, the formulæ of Brioschi appertaining hereto as well as our formulæ with the a_{ik}'s—to the single fact that the substitutions of the **A**'s can be co-ordinated to the even permutations of the x's simply and uniquely.* My reason for not entering upon these matters in the foregoing exposition is that I do not consider these wider speculations to which I have referred in I, 5 (§§ 4, 5, 9) as yet conclusive. I have the more readily confined myself to geometrical constructions of individual characteristics, believing that it is just by these that we shall be able to pass to a true insight into the general theory.

§ 9. On the Necessity of the Accessory Square Root.

We are at the end of our exposition; what we have yet to add concerns the necessity of that accessory square root which occurred in our first method in the Tschirnhausian transformation, in our second method, when we wished to effect the solution, of the problem of the **A**'s. We shall show that this square root is in fact indispensable if an icosahedral equation is to be reached at all; we shall further prove that from this follows that theorem of *Kronecker's* which we have mentioned in II, 1, § 7, and which declares the general impossibility of a rational resolvent with only one parameter for the ordinary equation of the fifth degree.

In order now to prove the first point, let us formulate our assertion as follows. Let $x_0, x_1, \ldots x_4$, be any five variable magnitudes, ϕ, ψ, two integral functions thereof without a

* "Ueber die Auflösung gewisser Gleichungen vom siebenten und achten Grade" (1879). See especially §§ 1–5. The mode of expression in the text supposes that to every permutation of the x's corresponds only one substitution of the **A**'s; single-valuedness in the reciprocal sense occurs also, but it would not be necessary for the success of the algebraical process.

common divisor. *Then it is impossible*, we assert, *to choose* ϕ, ψ, *in such wise that*

(29) $$\lambda = \frac{\phi(x_0 x_1 x_2 x_3 x_4)}{\psi(x_0 x_1 x_2 x_3 x_4)}$$

undergoes the icosahedral substitutions for the even permutations of the x's.

The proof presents itself at once if we consider that the original question, one belonging to the theory of functions, is transformed by virtue of the arbitrary choice of the x's into a question of the *theory of forms*. Namely, if, corresponding to any permutation of the x's, the substitution formula:

(30) $$\lambda' = \frac{\phi'}{\psi'} = \frac{\alpha\phi + \beta\psi}{\gamma\phi + \delta\psi}$$

were to occur, we could at once, on account of the arbitrary nature of the x's, write

(31) $$\phi' = C(\alpha\phi + \beta\psi), \quad \psi' = C(\gamma\phi + \delta\psi),$$

understanding by C an appropriate constant, so that, therefore, with the permutations of the x's, the two integral functions ϕ and ψ are transformed *bilinearly*. But now, as we showed in detail in I, 2, § 8, every group of binary substitutions which is to be isomorphic with the group of non-homogeneous icosahedral substitutions contains, of necessity, *more* than 60 operations, while to the 60 even permutations of the x's not more than 60 transformations of the integral rational functions ϕ, ψ, can correspond. This is an insurmountable contradiction, and therefore the method proposed in (29) is, in fact, proved to be impossible, *q.e.d.** The contradiction is not even removed if we now assume $\Sigma x = 0$, for every equation of the fifth degree can be transformed rationally into one with $\Sigma x = 0$.

For the sake of a better grasp of the essence of the proof, let us compare the theory of the canonical equations of the fifth degree. In them we have, besides $\Sigma x = 0$, $\Sigma x^2 = 0$ also; let us therefore write equation (30) as follows:

(32) $$\phi'(\gamma\phi + \delta\psi) = \psi'(\alpha\phi + \beta\psi),$$

* *Cf.* here and in the following paragraphs my oft-cited memoir in Bd. xii. of the Math. Annalen (1877), and also my communication to the Erlanger Societät of January 15, 1877.

then, in the case of canonical equations, it is by no means necessary that the two surfaces

$$\phi'(\gamma\phi+\delta\psi)=0, \quad \psi'(\alpha\phi+\beta\psi)=0$$

are identical with one another, but only *that they intersect in the same curve the canonical equation of the second degree represented by those conditions.* Now we have in any case decided that ϕ, ψ, and so likewise ϕ', ψ', have no common divisor. We shall also require that no factor shall be capable of being detached when we modify the functions which arise from the addition of proper multiples of Σx, Σx^2. Nevertheless, the curves of intersection of the canonical surface with $\phi'=0$, $\psi'=0$, may have a portion in common; this portion must be only an *incomplete curve of intersection,* and must not admit of being traced out by a surface appended to the canonical surface. If we assume that this is the case, no ground appears in fact for the existence of formula (31) (from which we deduced the contradiction). I must omit to work out in greater detail what I have said, and to show that in fact, in the reflexions thus given, our former treatment of the canonical equations of the fifth degree is absorbed. The proof which we have given of our primary assertion is extended without important modification to other cases also. First, we might substitute at once the problem of the **A**'s in place of the general equation of the fifth degree: we learn that it is impossible, in reducing this problem to an icosahedral equation, to dispense with the square root $\sqrt{A}$ (or an equivalent irrationality) as previously employed. We learn, further, that it is impossible to reduce the general equation of the fourth degree, by means of rational construction of resolvents, to an octahedral equation, or even, after adjunction of the square root of the discriminant, to a tetrahedral equation.* Moreover, we can now make a practical application

* As regards *equations of the fourth degree,* a solution can be effected in their case, as I here cursorily indicate, with the help of the octahedral equation (or of the tetrahedral equation), which is, so to say, a blending of the two methods, which for equations of the fifth degree are distinct. Denote as before the roots x_0, x_1, x_2, x_3, which are to be subject to the condition $\Sigma x=0$, by quadrilateral co-ordinates in the plane. Then we have the canonical conic $\Sigma x^2=0$, and we saw above (II, 3, § 2) how a point belonging to it can be determined directly by an octahedral equation or a tetrahedral equation. We shall now co-ordinate to an arbitrary point x of the plane a point y of the

of our train of thought. In this respect I, only remark that the property of $\mathbf{A}_0$, $\mathbf{A}_1$, $\mathbf{A}_2$, which was described just now (§ 5), may be deduced in the way thus indicated.

§ 10. Special Equations of the Fifth Degree which can be Rationally Reduced to an Icosahedral Equation.

We must now interrupt our general considerations, and make mention of special equations of the fifth degree which furnish an exception to the theorem just proved. In II, 2, § 4, we have given a geometrical interpretation of the resolvents of the fifth degree, and have seen that they can be represented by two half-regular twisted curves of deficiency zero. Our object now is to reverse this result. Let:

$$F(x, Z) = 0 \tag{33}$$

be an equation of the fifth degree with one parameter which admits of an interpretation of the kind mentioned. I assert that we can always reduce it by rational means to an icosahedral equation.

The proof is essentially the same as we have given in a somewhat different form in II, 3, § 1, in considering the canonical equation. By hypothesis the five roots of (33) admit of representation as rational functions of an auxiliary magnitude λ:

$$x_\nu = R_\nu(\lambda), \tag{34}$$

in such wise, that for appropriate variation of λ the x_ν's undergo any even permutation. We must now apply the proposition from the theory of rational curves, that this λ can always be

canonical conic as a covariant, by drawing from x the two possible tangents to the conic, and choosing one of the two points of contact. We can then establish the octahedral equation (or tetrahedral equation) on which y depends, and hence by inversion find x, &c., &c., all in strict analogy with the developments which we have opened up in the two concluding paragraphs of the preceding chapter.

In the case of *equations of the third degree*, all such prolixity, as we remarked in II, 3, § 2, disappears. In fact, we saw in I, 2, § 8, that the dihedral group of six substitutions, which comes under consideration in connection with them, can be very well transformed into the homogeneous form, without the number of its substitutions being increased; the grounds for the occurrence of the accessory irrationality which we have recognised as appropriate for equations of the fourth and fifth degrees are therefore wanting.

introduced as a rational function of the x's, and therefore in such wise that to every point of the curve corresponds only one λ.* I will assume, for the sake of brevity, that the λ appearing in (33) is already chosen in the manner here indicated. Then every one-valued transformation which transforms our curve into itself, in particular therefore every even permutation of the x_ν's, establishes for λ a one-valued transformation having a one-valued reciprocal and therefore a *linear* transformation. Thus we obtain corresponding to the 60 even permutations of the x_ν's a group of linear substitution, simply isomorphic with them, of the variable λ. By I, 5, § 2, this is of necessity the icosahedral group; it appears in the canonical form which we have always maintained as soon as we introduce in place of λ a proper linear function $\lambda' = \frac{a\lambda + b}{c\lambda + d}$ as parameter. *This* λ', *which is itself a rational function of the* x_ν*'s, then depends directly on an icosahedral equation, whereupon the proof of our assertion is accomplished.*

We append to what has been said a few stray remarks. First we see that we can reiterate our theorem with unimportant modifications in the problem of the **A**'s, or, if we like to take into consideration the octahedron or tetrahedron instead of the icosahedron, in the equation of the fourth degree. We recognise, further, that for the equation of the fifth degree there can be no rational twisted curves which for the whole of the permutations of the x_ν's pass over into themselves. Finally we remark that the occurrence of rational invariant curves (as we will express it) is altogether limited to those form-problems of which the group is simply isomorphic with one of the groups of linear substitutions of a variable which we have previously enumerated.

§ 11. Kronecker's Theorem.

We have now all the requisite materials for completing the proof of the oft-mentioned theorem of Kronecker. *Our object is to prove that it is impossible, in the case of any proposed equa-*

* *Cf.* the proof of this theorem in Lüroth's paper in Bd. ix. of the Mathematische Annalen (1875).

tion of the fifth degree, even after adjunction of the square root of the discriminant, to construct a rational resolvent which contains only one parameter.

Let us first remark that we can at once impart an apparently more precise formulation to this theorem, inasmuch as the group of the even permutations of five things is primitive.* Namely, we shall be able to derive, on the grounds stated, from every rational resolvent a fresh equation of the fifth degree $F(X)=0$ by means of renewed resolvent construction; and here we may at once subject the X's to the condition $\Sigma X=0$. The roots X_ν are here severally co-ordinated to the original x_ν's in such wise that the co-ordination remains unaltered for any even permutations of the x_ν's. We can therefore write as before:

(35) $$X_\nu = p \cdot x_\nu^{(1)} + q \cdot x_\nu^{(2)} + r \cdot x_\nu^{(3)} + s \cdot x_\nu^{(4)},$$

where $x_\nu^{(k)} = x_\nu^k - \frac{1}{5}\Sigma x^k$ and p, q, r, s depend rationally on the coefficients of the proposed equation and the square root of the corresponding discriminant. *All that we now have to show is this: that it is impossible to form from the general equation of the fifth degree, by means of a Tschirnhausian transformation* (35), *an equation of the fifth degree with only one parameter.*

To this end we must reflect generally as to what geometrical interpretation such an equation would have to receive. The totality of the arbitrary values x_0, x_1, x_2, x_3, x_4, form a simply connected continuum. If we therefore allow $x_0, x_1, \ldots x_4$, in (35) to alter arbitrarily, the point X will, at all events, trace out an *irreducible* locus. If we now add the supposition that the equation of the X_ν's contains only one parameter, the irreducible locus in question will have to be a *curve.* I say now *that the irreducible curve so obtained will be transformed into itself for the* 60 *even collineations of space.* In fact, in virtue of the convention which we have made concerning the coefficients p, q, r, s occurring in (35), the even permutations of the X_ν's correspond to the even permutations of the x_ν's; while, on the other hand, we can attain to every permutation of the x_ν's (and therefore in particular to every even permutation thereof) by allowing the x_ν's, beginning from any initial values, to move continuously in a suitable manner.

* *Cf.* the definition in I, 1, § 2.

We now return specially to the developments of the preceding chapter. Namely, it is evident *that the curve of the X_ν's just described must in every case be rational.* For we can suppose the $x_0, x_1, \ldots x_4$, in (35) rationally dependent in some way on a parameter λ, whereupon the X's themselves become rational functions of this λ: we need not regard the objection that in special cases the λ may altogether disappear from the X_ν's, since we can evidently always avoid such a contingency. The premises of the preceding paragraph are therefore in fact given. We conclude *that we can establish a rational function of the X_ν's which for the even permutations of the X_ν's undergoes the icosahedral substitutions.* This function would by virtue of (35) also depend rationally on the x_ν's in such wise that it would undergo icosahedral substitutions for the even permutations of the x_ν's. But now we have expressly proved in § 9 that such a rational function of the x_ν's is impossible. *We therefore arrive at a complete contradiction*, and must therefore give up our assumption that a Tschirnhausian transformation (35) exists with the property more precisely described above, *q.e.d.*

I conclude by adding a few more general observations on the theory of equations.

First, if in the foregoing exposition we substitute throughout the octahedron or tetrahedron for the icosahedron, we can repeat all our considerations unaltered for the equation of the *fourth degree* till we come to the one that treats of the primitivity of the corresponding group. The group of the equation of the fourth degree is composite. If, therefore, we wish to recover Kronecker's theorem for the equation of the fourth degree, we must expressly add to it the condition *that the group of the resolvent coming under consideration is to be simply isomorphic with the group of the twenty-four or the twelve permutations of x_0, x_1, x_2, x_3, x_4.* If we leave out this condition, rational resolvents of the general equation of the fourth degree may very well occur which contain only one parameter. The empirical proof thereof is effected by the ordinary solution of the equation of the fourth degree. In fact, this operates merely with auxiliary equations which contain only one parameter, namely, with binomial equations.

In the case of equations of the *third* degree, there can, of course, on the grounds of our previous remarks, be no question of a theorem corresponding to Kronecker's.

Concerning equations of a *higher degree*, I will here, in order not to be prolix, only make one remark, retaining therein, for the sake of simplification, the restriction which we formulated just now for the fourth degree. On the supposition mentioned, resolvents with only one parameter—disregarding special and easily-recognised cases—are impossible in the case of the general equation, for the reason that, according to the observation of § 10, among the corresponding invariant curves no rational ones can exist.

Lightning Source UK Ltd.
Milton Keynes UK
UKOW04f1447161015

260691UK00001B/42/P